TSUNAMI

To Survive from Tsunami

ADVANCED SERIES ON OCEAN ENGINEERING

Series Editor-in-Chief
Philip L- F Liu (*Cornell University*)

Published

*For the complete list of titles in this series, please write to the Publisher.

Advanced Series on Ocean Engineering — Volume 32

TSUNAMI

To Survive from Tsunami

Susumu Murata
Coastal Development Institute of Technology, Japan

Fumihiko Imamura
Tohoku University, Japan

Kazumasa Katoh
Musashi Institute of Technology, Japan

Yoshiaki Kawata
Kyoto University, Japan

Shigeo Takahashi
Port and Airport Research Institute, Japan

Tomotsuka Takayama
Kyoto University, Japan

NEW JERSEY • LONDON • SINGAPORE • BEIJING • SHANGHAI • HONG KONG • TAIPEI • CHENNAI

Published by

World Scientific Publishing Co. Pte. Ltd.

5 Toh Tuck Link, Singapore 596224

USA office: 27 Warren Street, Suite 401-402, Hackensack, NJ 07601

UK office: 57 Shelton Street, Covent Garden, London WC2H 9HE

British Library Cataloguing-in-Publication Data
A catalogue record for this book is available from the British Library.

First published 2010
Reprinted 2011

Advanced Series on Ocean Engineering — Vol. 32
TSUNAMI: TO SURVIVE FROM TSUNAMI

ISBN-13 978-981-4277-47-1
ISBN-10 981-4277-47-9
ISBN-13 978-981-4277-48-8 (pbk)
ISBN-10 981-4277-48-7 (pbk)

Printed by FuIsland Offset Printing (S) Pte Ltd. Singapore

Foreword

The world has recently experienced frequent happenings of earthquakes of historical magnitude with the accompanying giant tsunamis, which have brought about a great loss of human lives and precious assets. These include the Great Indian Ocean Tsunami (December 2004), the Chilean Tsunami (February 2010) and, most recently, the Giant Tsunami of Tohoku Pacific Coast in Japan (March 2011).

At 14:46 (local time) on 11 March 2011, a huge earthquake has occurred off the Pacific coast of the Tohoku region of northeastern Japan, with a magnitude rating of 9.0. This is the biggest earthquake since the national seismic survey started in Japan; a past earthquake of a similar magnitude in the region dates back to no less than 1100 years ago, according to old documents.

This earthquake was caused by a large fault, of width 200 km and length 500 km, under the seabed where the Pacific plate subducted the North American plate, causing the formation of a giant tsunami, with a height of 10-20 m. The tsunami totally devastated the entire 500 km of the northeastern Pacific coast of Japan. In many places, the tsunami also flooded over to the land areas, spanning several kilometers with a run-up height of 40 m above sea level and inundating a total area of 500 sq km.

The tsunami inflicted tremendous damage on the wide areas of northeastern Japan. Twenty-eight thousand people are either dead or missing, and two hundred and twenty thousand buildings are damaged (as of 9 April 2011). The total economic damage is estimated to be three hundred billion US dollars. Furthermore, a nuclear power plant located along the coast of Fukushima was also seriously damaged by this tsunami.

Thus, we observe that the disaster caused by this earthquake has proven to be an unprecedented wide-area disaster, which is now known as the "East Japan Great Earthquake Disaster".

Even for those of us who are far from the disaster sites, we can also sense the destructive power of tsunamis, and the unimaginable degree of damage which tsunamis inflict on man and nature in real time – through television, online images and other forms of the media.

We are reminded once again of the deadliness of a tsunami, which we have previously witnessed at the Indian Ocean Tsunami of 2004. In Japan, particularly at its northeastern Pacific coast, disaster prevention measures have been undertaken in great earnestness. Examples of these preventive measures include the use of hardware and software devices against the tsunamis, which are based on the experiences of frequent tsunami disasters throughout history. However, the tsunamis have – to date – completely overwhelmed all human efforts, destroying human lives, towns and villages in their rampage.

What we have witnessed through the media may have been unimaginable and unbelievable scenes for many in the world.

On the other hand, the Great Indian Ocean Tsunami – which has hit Indonesia and many countries within the region of the Indian Ocean – has left over three hundred thousand people dead or missing. These huge tsunami disasters suggest that earthquakes and tsunamis are already beyond human calculations. They may actually occur at any time in any coastal regions where submarine plates meet. This observation also carries an important implication that the mitigation of tsunami disaster has now become an important and imminent issue to be tackled by the world, and that the implementation of preventive countermeasures is urgently required.

In this book, the authors aim to give readers a virtual tsunami experience by reporting the findings of numerous on-site surveys immediately after the disasters, as well as providing detailed descriptions of various mechanisms related to tsunamis and the damage they cause. The authors also wish to educate readers on the ways to prepare for and deal with tsunamis, based on accurate, tsunami-related scientific knowledge.

In this book, the readers are invited to answer various questions on tsunami in the prologue. Hopefully, this will encourage them to gain a deeper understanding on the workings of the tsunamis.

The book also consists of two sections. "Part I: How to Survive a Tsunami" describes the characteristics of tsunamis and provides examples of the damages caused by them, while "Part II: Tsunami Behavior and Forecasting" contains the scientific descriptions of tsunami, for those readers who require more detailed knowledge on the subject matter.

This book has been made possible through the efforts of many. Leading researchers in the fields of tsunami and disaster prevention have served on the editorial committee and numerous researchers have also contributed to the articles. I am grateful to all of them for their cooperation in the preparation of this book.

In the face of the East Japan Great Earthquake Disaster, the authors have decided to include a new foreword to this book.

I would like to thank Mr. Yeow-Hwa Quek of World Scientific Publishing Co. Pte. Ltd. for his assistance and patience in this impression.

Susumu Murata, Committee Chairman
"*TSUNAMI: To Survive from Tsunami*" Editorial and Publication Committee
Coastal Development Institute of Technology
11 April 2011

Editorial Staffs and Authors

The editorial and publication committee invited leading researchers in the fields of tsunami and disaster prevention as contributors to the articles. We would like to introduce all contributors to the publication project as follow:

Editorial and Publication Committee of "TSUNAMI"
Coastal Development Institute of Technology

Committee Chairman	MURATA Susumu
Editorial Staff	IMAMURA Fumihiko KATOH Kazumasa KAWATA Yoshiaki TAKAHASHI Shigeo TAKAYAMA Tomotsuka

Authors

ARIKAWA Taro	Senior Researcher, Port and Airport Research Institute (1.2(5))
FUJIMA Kouji	Professor, National Defense Academy (4.3(1)(2))
HARADA Kenji	Research Associate, Saitama University (3.2)
HIRAISHI Tetsuya	Managing Executive, Port and Airport Research Institute (1.2(3), 2.4)
HUKAZAWA Yoshinobu	Counselor for Disaster Prevention, Fire and Disaster Management Agency (3.4)
IMAMURA Fumihiko	Professor, Tohoku University (1.4, 4.1, 5.1, 5.3)
KATOH Kazumasa	Visiting Professor, Musashi Institute of Technology (Prologue, Epilogue)
KAWATA Yoshiaki	Professor, Kyoto University (3.1)

KIMURA Katsutoshi	Professor, Muroran Institute of Technology (3.1)
KOSHIMURA Shun-ichi	Associate Professor, Tohoku University (3.4)
KUMAGAI Kentaro	Senior Researcher, Ministry of Land, Infrastructure, Transport and Tourism (1.2(4))
MAKI Norio	Associate Professor, Kyoto University (3.3)
MATSUTOMI Hideo	Professor, Akita University (1.5, 2.3)
SHIMOSAKO Kenichiro	General Manager, Ministry of Land, Infrastructure, Transport and Tourism (4.3(3))
TAKADA Masayuki	Director for Engineering Planning, Ministry of Land, Infrastructure, Transport and Tourism (3.5)
TAKAHASHI Shigeo	General Manager, Port and Airport Research Institute (1.1, 1.2(1), 2.2)
TAKAHASHI Tomoyuki	Associate Professor, Akita University (3.3)
TAKAYAMA Tomotsuka	Managing Director, Coastal Development Institute of Technology (2.1, 4.2)
TOMITA Takashi	Senior Researcher, Port and Airport Research Institute (1.2(2), 2.5)
YOMEYAMA Haruo	Senior Researcher, Port and Airport Research Institute (1.6)

(Authors are listed in alphabetical order and the numbers in parentheses show the sections in charge.)

Contents

Prologue

The Indian Ocean tsunami which occurred on December 26, 2004 caused unprecedented disaster and claimed the priceless lives of more than 300,000 persons worldwide. Among the conditions which contributed to the disaster were the enormous scale of the tsunami itself and inadequacies in the infrastructure, including tsunami disaster prevention facilities, evacuation facilities, emergency information communication systems, and others. However, it is regrettable to think that at least the loss of human life would not have been so extensive if proper awareness and scientific knowledge in connection with tsunamis had been more widespread and information on the tsunami had been transmitted quickly. The purpose of this book is to provide information and knowledge for tsunami survival to persons living in areas which have been or may be attacked by a tsunami, persons who may possibly live in such areas in the future, and travelers visiting such areas in the hope that even one more life may be saved if a tsunami attack occurs.

In virtually all cases, tsunamis are caused by submarine earthquakes, and these earthquakes bear no relation to the seasons, or time of day. Current technology has not clarified the phenomena that could give warning of an impending earthquake, and thus has not reached the stage where it is possible to predict earthquakes in advance. This of course means that tsunamis cannot be predicted. Furthermore, once a tsunami occurs, it travels to the coastline at almost same speed as a jet aircraft. How, then, can we survive when a tsunami occurs?

In fact, it is possible to survive a tsunami by understanding the experience and nature of tsunamis. Because this is discussed in detail in this book, only one example of the possibility of survival will be mentioned in this prologue. By nature, tsunamis travel faster in deeper water. Tsunamis travel at a speed of 800 km/h in ocean areas with a depth of 3,500 m, and at speed of 36 km/h in coastal areas of 10 m deep. In comparison with the speed with which seismic waves are transmitted (the speed of a S wave, which causes large tremors, is on the order of 12,600 km/h), the speed of tsunamis is actually very slow. This relative delay in the transmission of a tsunami gives us the chance to survive the tsunami. In simple terms, the seismic waves arrive at coastal regions almost simultaneously with the occurrence of an earthquake in an offshore ocean bottom, and the tsunami arrives there after some delay. This difference in arrival time is very small for the coastal region which is located in a seismic source area. Usually tsunami takes several to thirty minutes in its arrival after an earthquake is noticed.

This book explains, in concrete terms, the various methods of surviving a tsunami which are considered possible at present.

This book consists of two parts.

Part I examines the relationship of the human population with actual examples of tsunami disasters and the characteristics of tsunamis in the hope that even one additional person can survive a tsunami. First, Chapter 1 presents examples of tsunami disasters from the perspective of persons who were present at the scene of a tsunami disaster. While there are tsunamis which cause extremely great damage when they occur, their frequency of occurrence is extremely low. For this reason, the number of people who have actually experienced a tsunami is small, and the overwhelmingly large number of persons will never experience a tsunami in their lifetimes. Thus, the intention of this chapter is to give persons who have not experienced a tsunami a "quasi-experience" of a tsunami through concrete examples. Chapter 2 examines the engineering characteristics of tsunamis at the human scale. This is because we believe that it will be possible to take evacuation action while considering what may occur in the next instant if one understands the nature of tsunamis at the human scale (understands as knowledge "with the body"). In order to survive a tsunami, both "experience" and "knowledge" are necessary. "Experience" has an absolute strength which is not limited to tsunamis. However, it must be remembered that the same tsunami will not strike twice. This is because tsunamis with exactly the same natural and physical conditions do not strike, social conditions are different, in that the circumstances of coastal use change over time, and so on. Because the few available tsunami experiences contain a mixture of unique and general features, if the two are confused, "experience" may, on the contrary, become an impediment. What compensates for the drawback of unique features, which are difficult to eliminate from "experience," is universal knowledge. To survive a tsunami approaching in front of one's very eyes, it is necessary to continue to decide one's own actions almost instantaneously in response to changing circumstances based on experience and knowledge.

Chapters 1 and 2 present an explanation from the viewpoint of how to survive the terrifying experience of an actual approaching tsunami. In contrast, Chapter 3 describes preparations which can be made on an everyday basis and possible preparations against the attack of a tsunami. These are not preparations at the national level, but rather, focus on preparations which can be made at the individual, regional, and local levels. These routine preparations are also important for survival when an emergency occurs.

The second part, Part II, was prepared for engineers, persons involved in tsunami disaster measures which go beyond the individual level, and other persons who wish to obtain a more detailed knowledge of tsunamis. Chapter 4 provides a precise commentary on the characteristics and behavior of tsunamis. Chapter 2 discussed the nature of tsunamis from the viewpoint of survival. However, because Part I gives priority to simplicity and easy understanding, a detailed explanation was omitted, and this resulted in some lack of precision. Thus, Chapter 4 is intended to compensate for this. Chapter 5 treats the history of numerical simulation techniques for tsunamis and tsunami forecasting (warning) systems. Where numerical simulation techniques are concerned, the possible types and extent of simulations is explained for a technique (developed by Tohoku University in Japan) which has been certified as a world standard by UNESCO and can be used by anyone in any country. With regard to tsunami forecasting (warning) systems, this chapter introduces the history of systems in Japan, where various repeated efforts have gradually reduced the time from the occurrence of an earthquake to issuance of tsunami forecast (warning).

Finally, the authors would like to pose several questions regarding tsunamis. These were prepared referring to the inadequate understanding and misunderstanding of tsunamis which can be seen scattered through books and in the homepages and publications of public organizations. To what extent can you answer these questions?

A: The following explanations are incorrect. Where is the error?

(1) There are cases in which the ocean bottom near the epicenter of an earthquake repeatedly rises and falls due to movement of the earth's crust accompanying an ocean bottom earthquake. When this happens, large swells and depressions at the ocean surface propagate in all directions following the movement of the ocean bottom. This is the reason why a tsunami does not end with the first wave, but attacks repeatedly with 2nd and 3rd waves.

(2) The height of tsunamis in offshore areas is observed by weather observation ships. A microwave wave meter is attached to the bow of the weather observation ships owned by the weather bureau and marine observatories, and the distance from the sensor to the ocean surface is measured by emitting microwaves on the ocean surface. The height of the tsunami is then measured by eliminating the vertical movement of the ship itself.

(3) A tsunami is a type of wave. Because wave motion is basically circular, for example, if a diver is attacked by a tsunami while underwater at a depth of 30 m, the diver will be violently rolled from the bottom to the surface, creating an extremely dangerous condition. Furthermore, because the speed of the tsunami is several 10 km/h to several 100 km/h, it is impossible to keep a fixed position by grasping some object in the water.

(4) Immediately before a tsunami attacks, the water level near the coast line decreases. Accordingly, if the ocean bottom or rocks which one has not seen appear, attack by a tsunami is imminent, and one must immediately evacuate to higher ground.

(5) A tsunami warning was issued while you were on a certain small island. Fortunately, you were on the coast opposite the side where the

tsunami was approaching (i.e., on the shielded side). Therefore, you simply observed the condition of the ocean surface and prepared in case of emergency without making particular efforts to evacuate.

(6) The occurrence of a tsunami as such is unrelated to high and low tides, but when a tsunami propagates to a coastal area, the height of the tsunami will be greater if it occurs at high tide. For this reason, particular vigilance is necessary during high tides at the full moon and new moon.

(7) Numerous small waves occur over a wide range of the ocean surface above a submarine earthquake when such an earthquake occurs. These small waves gradually gather and overlap in the center as they approach land, and finally form an extremely large tsunami.

(8) The origin of the English word tsunami is Japanese. Among the Katsushika Hokusai's well-known 36 Views of Mt. Fuji, the "Great Wave off Kanagawa" is famous as a ukiyo-e depicting a tsunami and is frequently reproduced in books on tsunami in other countries.

B: What are the grounds for the following five rules regarding tsunamis? (Seven Ministry Liaison Council on Tsunami Countermeasures, 1997)

(1) If you feel a strong earthquake (seismic intensity 4 or higher) or slow tremors over an extended time during a weak earthquake, immediately leave the seashore and quickly evacuate to a safe location.

(2) Even if you do not feel the earthquake, when a tsunami warning is announced, immediately leave the seashore and quickly evacuate to a safe location.

(3) Obtain correct information from the radio, television, information broadcast vehicle, etc.

(4) Even in case of a tsunami alert (tsunami watch), ocean swimming and beach fishing are dangerous and must be avoided.

(5) Because tsunamis attack repeatedly, do not relax your vigilance until the warning or alert is lifted.

How did you do with these questions? Even if you are not able to answer them now, we hope that you will remember them and ask yourself whether you can answer them all when you finishing reading the book.

Part I

How to Survive a Tsunami

Chapter 1

Examples of Tsunamis and Tsunami Disasters

1.1 Case Studies of Actual Tsunami Disasters

Figure 1.1 shows the tsunami approaching the coast of Thailand and people trying to escape from the tsunami during the Great Indian Ocean Tsunami, which occurred in December 2004. The people in the photograph are trying to escape from it, realizing that the danger is serious. Fortunately, all of these people survived, but did they really understand what was happening at the time, what would follow this scene, and what they should do?

Fig. 1.1 Swimmers fleeing an attacking tsunami (Thailand, 2004).

In order to plan for a smooth evacuation and take effective disaster prevention measures, it is important to understand tsunamis. In particular, each individual must understand the actual damage that tsunamis can cause. For example, if the people in the photograph had understood tsunamis, they would have realized the danger of a tsunami when the sea initially retreated. And if they had understood the damage caused by tsunamis, they would have moved quickly to safety. Even after a tsunami begins to attack, the possibility of survival can be increased by escaping to a nearby high building or other safe place. In other words, an understanding of tsunamis makes it possible to take proper action to avoid danger, and thereby increases the possibility of survival.

Unfortunately, even residents of tsunami-prone regions rarely have direct experience of a tsunami and do not understand the real nature of tsunami disasters. When a tsunami occurs, it can cause terrible damage over a wide region, including heavy loss of human life, as seen in the Great Indian Ocean Tsunami. However, catastrophic tsunamis are very rare. What's more, people from countries which are not prone to tsunami attack (like those in Fig. 1.1) may not even know the word "tsunami." For this reason, a simulated "quasi-experience" of a tsunami is important, for even this indirect experience can be as effective as a vaccination for preventing disease.

Figure 1.2 shows video images taken in Sri Lanka during the same Great Indian Ocean Tsunami. These images were taken at the bus terminal in the center of Galle, Sri Lanka. The tsunami caused an extremely violent flow in the bus terminal from the left rear. The video shows cars and the remains of houses being carried along by the tsunami and people caught in the current. Images like these were broadcast around the world, and many people saw the terror of tsunamis with their own eyes. In other words, seeing video images of a tsunami is a valuable experience for disaster prevention, even for people far from the actual scene.

However, we must remember that images can only show a very small part of a tsunami disaster. In many cases, TV images show only shocking scenes, and not the whole picture of the tsunami. Explanations of images are also inadequate. Tsunami disasters are quite different, depending on the actual tsunami and the conditions in the area. This means that the tsunamis assumed in each region and the damage they cause will not necessarily be the same as shown in television images. Images of tsunami disasters are valuable and extremely effective in educating people, but it is dangerous to depend only on those images. Therefore, this chapter aims to give the reader a more accurate understanding of tsunamis by describing the real nature of a tsunami disaster, as concretely and scientifically as possible, based on the results of scientific study. The nature and severity of tsunami disasters are determined by the conditions related to the attacking tsunami and the attacked region. The most basic parameter of a tsunami is its height. The

Fig. 1.2 Violent flow caused by an attacking tsunami (Sri Lanka, TV).

basic parameters of the region are its distance from the coast and height above sea level, as well as those of coastal disaster prevention facilities (for example, breakwaters and tsunami control forests), and buildings (type construction, density), and the state of human activities (sleeping, working, commuting to work). While considering these factors, we will try to make the reader feel the reality of tsunamis from the human perspective, utilizing photographs where possible.

When reading each section, the reader should remember that the explanation is given for a certain height of tsunami. This is done to clarify the relationship between the height of the tsunami and the nature and degree of damage, because the magnitude of a tsunami is represented by its height. Therefore, if you can estimate the height of a tsunami attacking your region, you can form a specific image of the possible damage in your area by referring to the examples in each section. Forecasts of tsunamis for alerts and warnings currently use the height of the tsunami as a standard, as shown in Table 1.1, that is, warnings say "an *x*-meter tsunami is forecast." It is possible to imagine how large a tsunami attack will be and how much damage will occur by comparing with the damage examples in each section, and to predict expected damage and make evacuation plans based on this information.

Table 1.1 Standards for tsunami warnings and alerts.

Type of forecast	Typical announcement	Height of tsunami (in meters)
Tsunami alert	A tsunami with a maximum height of 0.5 m is forecast; caution is advised.	0.5
Tsunami warning	A tsunami with a maximum height of 2 m is forecast; extreme caution is advised.	1 or 2
Large tsunami warning	A tsunami with a maximum height of 3 m or higher is forecast; extreme caution is advised.	3, 4, 6, 8, 10, or higher

1.2 Indian Ocean Tsunami Disaster

(1) *Outline of the earthquake and tsunami, and resultant disaster*

(1) *Occurrence of the earthquake and tsunami*

An earthquake with a magnitude of *M*9.3* occurred on December 26, 2004 at 0:58 GMT (Greenwich Mean Time; 7:58 in the morning in the local time of Indonesia and Thailand, and 6:58 in Sri Lanka). The occurance of the earthquake was captured by seismographs around the world, and general estimates of the focus and magnitude were given immediately by meteorological agencies in Japan, the United States, and other countries. The focus was the sea bottom (latitude 3° N, longitude 96° E, depth 30 km) approximately 10 km northeast of Pulau Simeulue (Simeulue Island) off the coast of the Island of Sumatra, Indonesia. The rupture of the earth's crust which caused the earthquake propagated to the Nicobar and Andaman Islands to the north, reaching a total length of about 1,000 km. This was a trench-type earthquake, which causes large-scale tsunamis. It occurred in the earthquake-prone region where the Indo-Australian Plate subducts under the Eurasian Plate. The magnitude was extremely large, even from a global viewpoint, and was the fourth largest since 1900 (the largest was the *M*9.5 Chilean Earthquake of 1960).

(2) *Propagation of the tsunami*

In order to understand the disaster caused by this tsunami, first, it is necessary to know how it spread from the focal area, propagating through the open sea, and attacked the coasts of countries around the Indian Ocean. The propagation of the tsunami can be seen in numerical simulations by a computer. Figure 1.3 shows the results of such simulations. (a) shows the estimated location where the sea bottom ground rose and fell due to the rupture of the earth's crust. The areas in

---

* The Geological Survey and North-Western University in USA proposed *M*9.1 and *M*9.3, respectively, as the seismic magnitude.

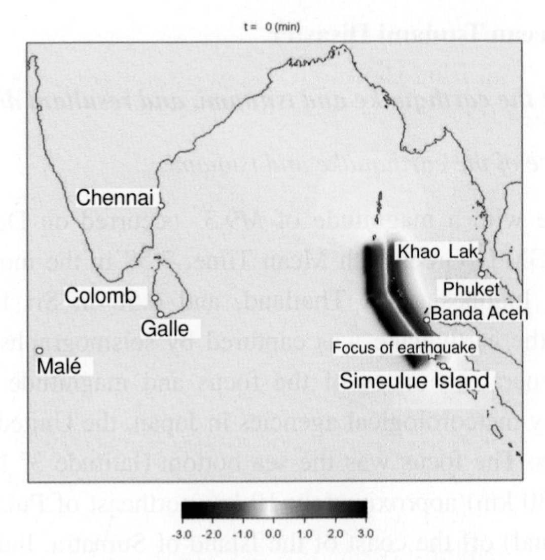

(a) Initial displacement of sea bottom ground

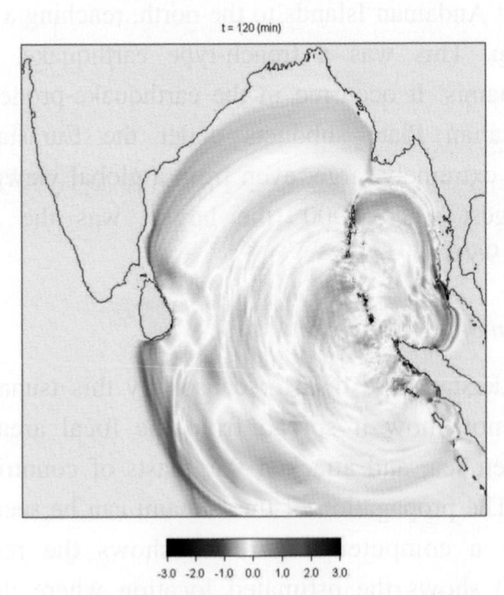

(b) Propagation of the tsunami (after two hours)

Fig. 1.3 Example of numerical simulation of the Great Indian Ocean Tsunami (Tomita et al., 2005).

red are where the seabed rose. The average height of the rise was approximately 2 m. Blue shows areas that sank, the average drop being from -0.5 m to -1.0 m. Banda Aceh, which is on Sumatra near the focus of the earthquake, sank. This deformation of the seafloor shape was transmitted unchanged to the sea surface, where it caused a corresponding deformation that resulted in the tsunami. Figure 1.3(b) shows the tsunami two hours after the earthquake. Because the energy of a tsunami spreads and weakens with distance, the height of the tsunami decreases as it propagates over long distances. The direction and height of the tsunami also change due to the topography of the sea bottom, and the tsunami becomes more complex due to reflection by land and islands. In the open sea, the height of a tsunami is normally less than 2 m. Because this is small in comparison with its length that reaches several 100 kilometers, it is difficult for human observers to recognize a tsunami, and the effect on ships is extremely slight. However, after a tsunami propagates across the open sea, its height increases when it approaches the coast as the water becomes very shallow. As a result, the tsunami causes damage on the coastline. A tsunami is also affected by the topography of the sea floor (through various effects such as refraction, diffraction, and energy concentration), and the wave height and direction vary greatly, depending on the place. Damage also differs, depending on these conditions.

(3) *Tsunami attack*

At Banda Aceh in northern Indonesia, near the focus of the earthquake, the tsunami attacked less than 20 minutes after the earthquake. The height of the tsunami (trace height or run-up height: a maximum height of a tsunami onshore above sea level) was 4-8 m in districts facing north, but was as large as 10-20 m in districts on the west, which faced the focus of the earthquake. The average water depth of the ocean between the seismic focus and Thailand is relatively shallow, at approximately 400 m. As a result, the propagation velocity of the tsunami was also comparatively slow, being on the order of 200 km/h. In Thailand, the tsunami run-up height was 4-10 m in the vicinity of Khaolak and further to the south, 4-6 m around Phuket. A notable point is that because the sea bottom ground facing Thailand dropped, the tsunami attack on the Thai coast began with a "backwash" (water pulling back from the normal shoreline). On the other hand, the ocean depth between the focus of the earthquake and Sri Lanka was deep, averaging approximately 4,000 m. As a result, the speed of the tsunami was extremely fast, at about 700 km/h, and the tsunami propagated over a distance of about 1,200 km in a little more than 100 minutes. The fundamental period of this tsunami was 40-50 minutes, and its wavelength was several 100 km. In Sri Lanka, the height of the tsunami was 5-10 m on the eastern coast and 4-10 m on the southern coast. Because the sea bottom ground around the earthquake focus facing Sri Lanka rose, the tsunami attack on the Sri Lanka coast began with a rising wave. The tsunami also propagated to India and the islands of the Maldives, and even reached far-away Somalian coast in Africa causing damage. In these regions, the tsunami attacked repeatedly, and in many cases, not the 1st wave, but the 2nd or later wave was the largest. However, in Bangladesh, which is located along the axis of the displacement of the sea bottom ground, the height of the tsunami was not particularly great.

(4) *Human damage due to the tsunami*

The Indian Ocean Tsunami resulted in a very large number of deaths and missing persons (Table 1.2), totaling around 300,000 persons. In general, a disaster which causes around 10,000 deaths is called a "catastrophic disaster." The Indian Ocean Tsunami far exceeded this standard, and was one of the worst disasters in recent history. The following three reasons can be given for this enormity of the disaster.

(1) A giant earthquake caused an extremely large tsunami which attacked many regions, including remote countries. Because the earthquake tsunami was caused by a giant trench-type earthquake, a large tsunami with a height exceeded 4 m attacked many regions. Heavy damage, including destruction of buildings occurred in areas where the tsunami exceeded 4 m. Areas where the tsunami was 8 m to over 10 m were basically wiped out.

Table 1.2 Deaths and missing by country.

Indonesia	256,000
Thailand	5,400
Sri Lanka	38,000
India	10,750
Maldives	80
Somalia	300

(2) The tsunami suddenly attacked the people who did not know that a large earthquake had occurred and also did not receive any warning in advance. Collapse of buildings and other earthquake damage occurred in the western and northern parts of Sumatra Island in Indonesia and neighboring areas near the focus of the earthquake as well as in the Andaman Islands of India. However, there was no earthquake damage in other countries, and the tremors in Thailand were very slight. In these countries, the earthquake tremors were not felt; in addition, there was no warning system. As a result, no one realized that an earthquake had occurred; no one was aware of the possibility of a tsunami attack in advance, resulting in a very large disaster.

(3) In these regions, many people live on low-lying land near the coast, and there were almost no coastal disaster prevention facilities. In the regions where this tsunami disaster occurred, the effect of cyclones is not particularly serious and the ocean is relatively calm. Because of these conditions, many people had lived in relatively simple houses near the shore.

In particular, (2) and (3) are very different from the conditions in Japan, where tsunamis occur with comparatively high frequency and disasters due to storm surge in typhoons are common.

In addition, because Thailand has many beach resorts, the victims included a large number of tourists. Table 1.3 shows the number of deaths and missing tourists by country. The fact that many tourists who knew nothing about tsunamis died is one distinctive feature of this tsunami disaster.

Table 1.3 Deaths of tourists by country.

Country of origin	Dead and missing (presumed dead)	Country of origin	Dead and missing (presumed dead)
Germany	619	Denmark	47
Sweden	575	Australia	41
Great Britain	248	Hong Kong	40
Italy	210	Netherlands	39
New Zealand	209	Canada	36
Finland	189	South Korea	20
United States	169	China	18
Switzerland	157	Ukraine	17
Austria	114	Philippines	15
France	96	South Africa	15
Japan	93	Poland	11
Norway	84	Others	110

References

Tomita, T., Honda, K., Sugano, T., and Arikawa, T. (2005): Field Investigation on Damages due to 2004 Indian Ocean Tsunami in Sri Lanka, Maldives and Indonesia with Tsunami Simulation, Technical Note of the Port and Airport Research Institute. (in Japanese)

The Investigation Delegation of the Japanese Government on the Disaster Caused by the Major Earthquake Off the Coast of Sumatra and Tsunami in the Indian Ocean (2005): Great Indian Ocean Tsunami Disaster of December 26, 2004 – Report of Investigation – Cabinet Office, p. 174. (in Japanese)

(2) *Disaster in Banda Aceh, Indonesia*

(1) *Outline of Banda Aceh*

Minutes before 8:00 in the morning on December 26, 2004, a giant earthquake *M*9.3 occurred in the ocean approximately 100 km west of the coast of Sumatra Island. The epicenter of this earthquake was 250 km south-south-east of the city of Banda Aceh (Fig. 1.4). However, the focal region then spread 1,000 km to the north from the original focus, and the sea bottom ground was moved by the earthquake even in the area to the west of Banda Aceh. The tsunami that accompanied this ground movement attacked Banda Aceh. According to an eyewitness who observed the tsunami from a high ground on the coast west of Banda Aceh, the tsunami attack occurred about 15-20 minutes after the tremors caused by the earthquake were felt. In some places, the tsunami that attacked the shore was higher than 10 m. At the time, Banda Aceh had a population of 264,618 persons. Of these, 61,065 died (source: Banda Aceh City homepage).

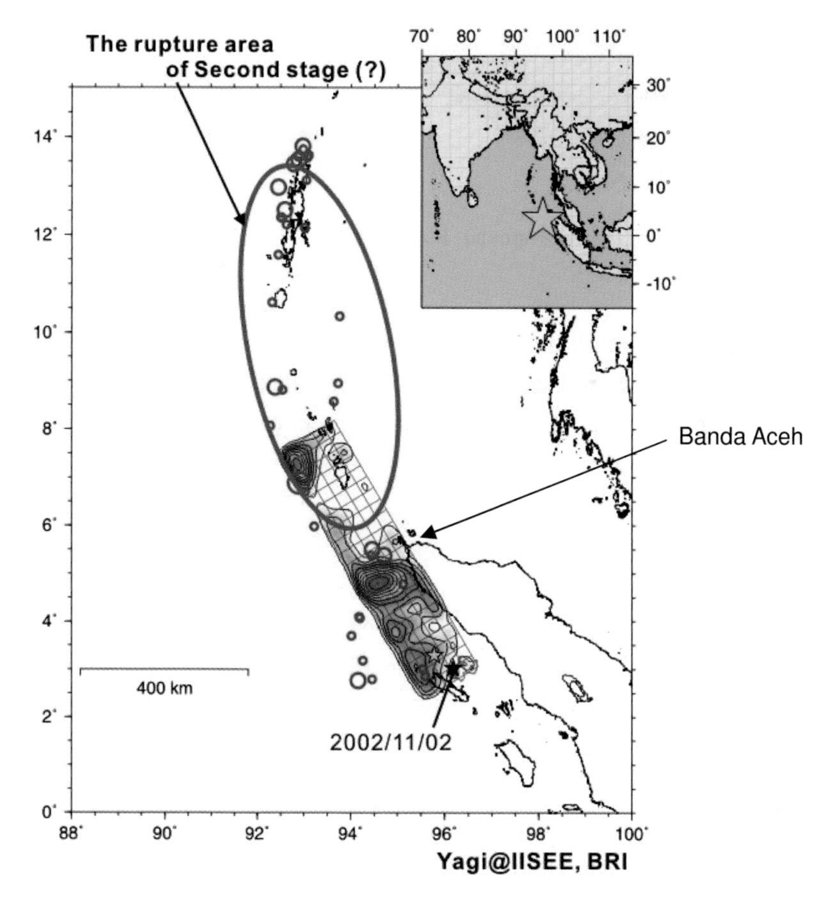

Fig. 1.4 Results of an estimation of the position of the focus of the earthquake and the amount of slip of the fault (Yagi, 2004). The star mark in the figure shows the position of the focus; large slipping of the fault occurred in the areas marked in red.

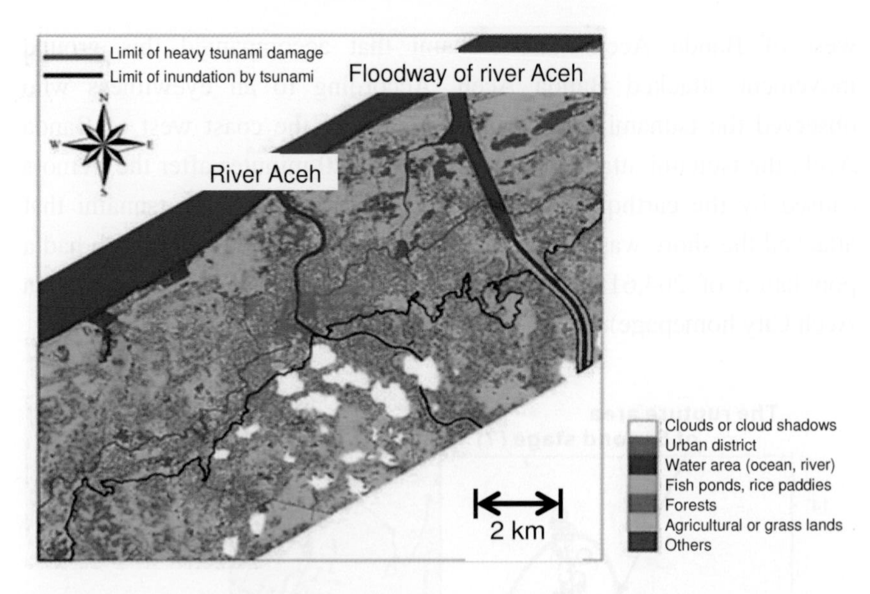

Fig. 1.5 Land use in the area around Banda Aceh (Geographical Survey Institute, Government of Japan, 2004).

Banda Aceh is built on flat land around the mouth of the Aceh River. In the hinterland beyond the coastline, fish ponds and rice paddies cover the area of 11 km^2 (Fig. 1.5). The center of the city is located approximately 2.5 km inland from the coast. However, the height of the ground was about 2 m above sea level at the time of the tsunami attack, and the low flat land extends from the coast to the inland.

(2) *Attacking tsunami and inundation*

The height (trace) of the attacking tsunami was 12 m at a mosque near the coast. The trace of the tsunami was found at a height of 7 m even at the mouth of the Aceh River to the east of this structure. This giant tsunami caused devastating damage 2-3 km far from the coast and inundation up to 5 km from the coast (Fig. 1.6).

At a high school in an area (marked by the circle in Fig. 1.6) 2.5 km far from the coast, the concrete school building remained after the tsunami, and water marks were left on its walls. The ground height in this area is 1.3-1.6 m above sea level, and the water marks of the tsunami on the side walls of the concrete building were at a height of 4.0-4.4 m above the ground (Fig. 1.7). In other words, this flat, low-lying city was destroyed by water flows that exceeded 4 m in height (Fig. 1.8).

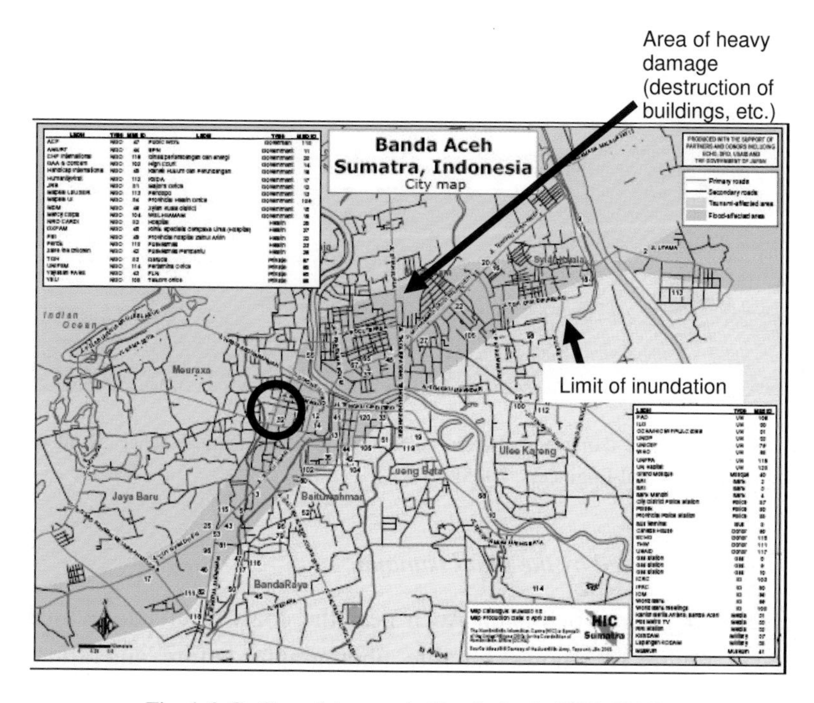

Fig. 1.6 Outline of damage in Banda Aceh (HIC, 2005).

Fig. 1.7 Water marks left on the side wall of a school 2.5 km far from the coast (height above ground: 4.4 m).

Fig. 1.8 View of the damage in Banda Aceh (US Navy, Reconnaissance Team of the Japan Society of Civil Engineers, 2005).

(3) *Damage caused by the giant tsunami*

The residential area of Banda Aceh near the coast received devastating damage. Figure 1.9 shows satellite photographs before and after the tsunami. Although these are difficult to see in the photo, erosion countermeasures had been constructed on the shore of the sandbar facing the sea in order to protect the coast from storm surge. However, these countermeasure works were damaged by the tsunami. As a result, the land behind the works was eroded and the coastline was moved back. The land in the center of the sandbar was completely submerged as a result of erosion. As these photographs show, a tsunami has the power to change the shape of the land.

Figure 1.10 shows the remains of a residential area on the sandbar in Fig. 1.9(a) after the tsunami. Attacked by a tsunami as high as 10 m, houses were easily swept away, and even the pavement of roads was ripped up and carried away. There is a high possibility of damage to reinforced concrete (RC) buildings when the height of a tsunami exceeds

(a) Before the tsunami

(b) After the tsunami

Fig. 1.9 Sandbar eroded by the tsunami (© Digital Globe).

Fig. 1.10 Remains of destruction caused by the tsunami.

Fig. 1.11 Ferry terminal building under construction, which survived the tsunami.

5 m. However, one RC building on this sandbar withstood the tsunami. This was the ferry terminal building (Fig. 1.11). The first floor collapsed, but the second and higher floors remained. It may be noted that the first floor was destroyed by the earthquake, and not by the tsunami. This terminal building was under construction at the time. The framework of RC columns had been completed, but the walls had not been constructed. The 10-m tsunami attacked the building at this stage of construction and it passed cleanly through the building because it happened to have a wall-less "piloti" structure. As a result, the building escaped destruction by the tsunami. ("Pilotis" are piers. In a piloti structure, the building is raised above the ground on piers. In other words, the first floor consists of piers and open space without walls.) This example shows that a piloti structure can be effective as tsunami countermeasures, provided the structure is strong enough to withstand the earthquake ground motion.

The high tsunami also pushed and moved gigantic objects which would normally seem impossible. Figure 1.12 shows a power generating barge that was carried about 3 km inland from the coast by the tsunami. The barge is approximately 60 m long and 20 m wide. This barge was moored in the port before the tsunami. The water marks left by the

Fig. 1.12 Power generating barge carried onshore by the tsunami.

tsunami on buildings near where the barge was moved were about 3 m above the ground. This is roughly the same as the draft of the barge. It seems that the barge finally stopped when its bottom stuck on the ground. Large ships were carried inland in other areas as well, and these were also flat-bottomed boats. This shows that flat-bottomed boats are easily run up on land by tsunamis.

In Kreung Raya Port to the east of Banda Aceh, the tsunami reached 5 m above ground level. Here, three empty oil tanks (out of a total of nine tanks) were carried by the tsunami. These tanks, which were 17 m in diameter and 11 m in height, were carried about 300 m along the coastline (Fig. 1.13).

At a cement works on the west coast of Sumatra, an oil tank was heavily deformed by the force of the tsunami (Fig. 1.14). The height of the tsunami was estimated at more than 20 m. The destructive force of the tsunami can be imagined from this picture.

Fig. 1.13 Oil tank carried by the tsunami.

Fig. 1.14 Oil tank heavily deformed by the tsunami.

Fig. 1.15 Building after the first floor collapsed in the earthquake.

(4) *Large tremors caused by earthquake*

Before the tsunami attack, the city of Banda Aceh was also hit by large tremors (earthquake ground motion). On the Japanese scale, the earthquake intensity was a little less than 5 to a less more than 6. This intensity indicates that the tremors were so large as to make it difficult for a person to keep standing on the ground. Many RC structures were destroyed by this shaking. As shown in Fig. 1.15, several of these were destroyed when the first floor (ground floor) collapsed. As mentioned previously, Banda Aceh is on flat land with no high ground where people can escape from a tsunami. In flat areas like this, high buildings are essential as places of refuge from tsunamis. Thus, buildings which can withstand earthquakes (that is, earthquake resistant buildings) are important.

(5) *Condition of the tsunami attack*

When the focus of an earthquake is nearby, people feel large tremors, and this can trigger an evacuation from a tsunami. Unfortunately, however, people do not always evacuate. For example, the earthquake was extremely strong in Banda Aceh, but the residents had no previous experience of tsunamis. As a result, nobody realized that a tsunami might follow the earthquake, and nobody evacuated.

The tsunami attacked Banda Aceh about 15 minutes after the giant earthquake. According to witnesses, first, the ocean pulled back toward offshore, and then the tsunami attacked like a wall of water. Similarly, on the coast west of Banda Aceh, witnesses said that first the ocean pulled back about 2 km from the normal coastline. Thus, it seems that the tsunami that attacked Sumatra was a "backwash-first" type tsunami, which began with the water pulling away from the shore. Following this, many people went out onto the exposed sea bottom to collect fish, not knowing anything about an imminent tsunami attack on land, and it is said that some of these people fell victims to the tsunami.

The tsunami attack at Banda Aceh did not end with one wave. Witnesses stated that the tsunami attacked three times, and the 2nd wave that attacked about 15 minutes after the 1st wave was largest. According to survivors from the western side of the city, the 1st wave came from the north side that directly faced the coast, but the stronger 2nd wave attacked from the west. The west side of the city does not directly face the coast. However, there is a creek that flows to the north coast, and it seems that the tsunami ran up by this route. On the other hand, on the northern coast, the 1st wave of the tsunami attacked almost directly from the north, but the 2nd wave attacked somewhat more from the west. These facts suggest that the tsunami that flood the land displayed complex behavior because of the effect of the land topography (including the creek) and the direction of the tsunami attack from offshore area.

(6) *Evacuation from the tsunami*

Because it is generally said that tsunamis travel at a speed of around 700 km/h, some people believe that it is impossible to escape after seeing a tsunami. However, 700 km/h is the speed of a tsunami traveling across open ocean with a depth of 4,000 m (average depth of the Pacific Ocean). The speed of a tsunami depends on the water depth, becoming slower as the water depth becomes shallower. Therefore, do not give up hope even if you actually see a tsunami approaching. Even in areas where the maximum height of the tsunami on land was more than 4 m, some people escaped by running survived. He said he saw black water approaching from the north and escaped by running to the south. Video images of the tsunami in the town show many people running from the leading edge of the tsunami and escaping to buildings or other high places (Fig. 1.16). According to an analysis of these video images, the maximum inundation depth was approximately 1.6 m and the velocity of the tsunami was around 5 m/s, or 18 km/h. Thus, the lower the water depth, the slower the speed of the tsunami flow. Therefore, if you are so unfortunate as to see a tsunami running toward you, it is important not to give up hope, and to escape to some high place.

Fig. 1.16 People fleeing the leading edge of the tsunami in Banda Aceh.

(a) Mosque, as seen from the sea

(c) Hole used to escape onto the roof

(b) View from inside the mosque

Fig. 1.17 Mosque by the sea, which was attacked by a 12-m tsunami.

Figure 1.17 shows a mosque near the coast (Baiturrahim Mosque; "Mosque by the Sea"), which was attacked by a tsunami of 12 m in height. Many mosques are constructed more strongly than ordinary houses, and survived both the earthquake and the tsunami. People also survived by fleeing to this mosque. However, because the 12-m tsunami reached just below the roof of the mosque, people broke through a wall leading to the roof and climbed onto the roof in order to escape.

Other people escaped by car. One person saw the approaching tsunami from afar, jumped in his car, and drove quickly along a straight road in front of his house. When he turned a corner onto a larger street, the tsunami was already there. This highlights the fact that wide roads become “channels” for a tsunami, and the tsunami travels faster on these roads than in areas with buildings and other structures. This example should also make us think about how to evacuate safely.

As the tsunami moved through the town of Banda Aceh, it destroyed many houses and other buildings. The flow of the tsunami included debris from these destroyed buildings, mud, floating automobiles, and other objects (Fig. 1.18). The color of the water was dirty and muddy. The flow also carried large ships, fishing boats, etc. (Fig. 1.19). In addition to drowning, there are following many other dangers if you are caught in this kind of tsunami. You may be thrown against a building or rock, you may not be able to keep your head above water, or you may be injured by floating objects. There are also several examples of lucky people who held onto a floating object and were rescued offshore. However, this is generally very difficult. Even if you are confident about your swimming ability, it is important to flee before you are caught in a tsunami.

Fig. 1.18 Debris carried by the tsunami.

Fig. 1.19 Fishing boat carried onshore by the tsunami.

References

Geographical Survey Institute, Government of Japan (2004): Grasp of tsunami disaster conditions using high resolution satellite data (Banda Aceh, Indonesia): http://www1.gsi.go.jp/geowww/EODAS/banda_ache/banda_aceh.html (date referenced: June 28, 2007)

Sakakiyama, T., Matsutomi, H., Tsuji, Y., and Murakami, Y. (2005): Comparison of tsunami flood flow velocity by video images and estimated values from site survey, Tsunami Engineering Technical Report, Vol. 22, pp. 111-118. (in Japanese)

Shuto, N. (1992): Tsunami intensity and damage, Tsunami Engineering Technical Report, Vol. 9, pp. 101-136. (in Japanese)

Tomita, T., Honda, K., Sugano, T., and Arikawa, T. (2005): Field Investigation on Damages due to 2004 Indian Ocean Tsunami in Sri Lanka, Maldives and Indonesia with Tsunami Simulation, Technical Note of the Port and Airport Research Institute, No. 1110, p. 36. (in Japanese)

Banda Aceh city homepage: http://www.bandaaceh.go.id/index.htm (date referenced: June 28, 2007)

Reconnaissance Team of Japan Society of Civil Engineers (2005): The Damage Induced by Sumatra Earthquake and Associated Tsunami of December 26, 2004, A Report of the Reconnaissance Team of Japan Society of Civil Engineers, p. 97. HIC (2005): Banda Aceh Indonesia City, Map http://www.humanitarianinfo.org./Sumatra/mapcentre/ListMaps.asp?type=thematic&loc=City%20Map (date referenced: June 28, 2007)

Tsuji, Y., Matsutomi, H., Tanioka, Y., Nishimura, Y., Sakakiyama, T., Kamataki, T., Murakami, Y., Moore, A., and Gelfenbanm, G. (2005): Distribution of the Tsunami Heights of the 2004 Sumatra Tsunami in Banda Aceh Measured by the Tsunami Survey Team (Head: Dr. Tsuji): http://www.eri.u-tokyo.ac.jp/namegaya/sumatera/surveylog/eindex/htm (date referenced: June 28, 2007)

USGS (2007): Magnitude 9.1 – Off the West Coast of Northern Sumatra: http://earthquake.usgs.gov/eqcenter/eqinthenews/2004/usslav/ (date referenced: June 28, 2007)

Yagi, Y. (2004): Preliminary Results of Rupture Process for 2004 Off Coast of Northern Sumatra Giant Earthquake (Ver. 1): http://iisee.kenken.go.jp/staff/yagi/eq/Sumatra2004/Sumatra2004.html (date referenced: June 28, 2007)

(3) *Disaster in the Phuket Island/Khao Lak area, Thailand*

Southern Thailand is a resort area centering around Phuket Island. Tourism-related development has also progressed on the Khao Lak coast on the mainland side. As a result, many resort hotels are located along the coast. December 26, 2004, when the tsunami struck, was the height of the tourist season in the midst of the Christmas and New Year's holidays, and large numbers of tourists were visiting these areas. The death toll in southern Thailand was approximately 5,400 persons, including more than 2,000 foreign tourists who were confirmed dead. Many of the dead were from northern European countries (Sweden: 543, Germany: 537, Finland: 179, UK: 149, France: 95, Denmark: 45, Norway: 84, Austria: 86, and others). Looking at these numbers, one receives the impression that the victims tended to be from countries with almost no experience of earthquakes or tsunamis. However, many Japanese also died, even though Japan is an earthquake-prone country which has experienced repeated tsunamis.

The earthquake occurred in western Indonesia on December 26 at 7:58 a.m., Thai time. The tsunami attacked Phuket Island and the Khao Lak coast about two hours later. On Phuket Island, the trace height of the tsunami relative to sea level was approximately 5 m at Patong Beach (ground height: approx. 3.5 m). On the Khao Lak coast (ground height: approx. 2.5 m), the trace height was approximately 10 m, and at the fishing port of Baan Nam Khem on the northern Khao Lak coast, the trace height was approximately 6 m (Fig. 1.20). On the Khao Lak coast, where the height of the attacking tsunami was large, the ground height was low, the topography of the hinterland behind the coast was flat, and low ground extended inland. As a result of these factors, inundation by the tsunami spread as far as about 1 km inland.

At around 7:00 in the morning, a staff member of a restaurant located at Patong Beach, in Phuket Island, felt the earthquake while he was at home in Phuket Town (a town about 15 km from Patong Beach; the town itself only suffered slight damage from the tsunami). He saw the surface of the water in the bath tub sloshing and knew that an earthquake had occurred. However, he did not receive any information about where it had occurred or whether there was a possibility of a tsunami attack. Not worrying about the possible danger, he went to Patong Beach to open the restaurant and was there when the tsunami attacked about two hours later.

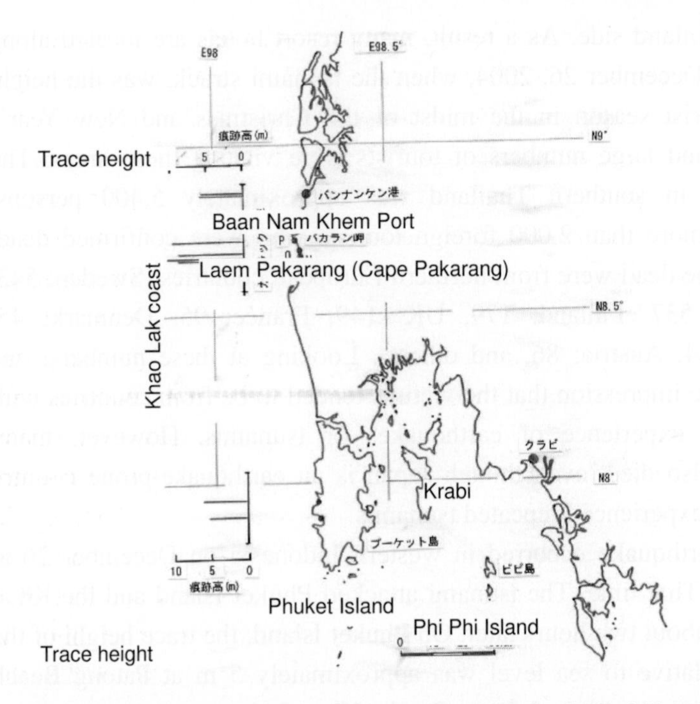

Fig. 1.20 Distribution of trace height in southern Thailand (Hiraishi et al., 2005).

A Japanese man named Mr. Honda, who also lived in Phuket Town, recognized the earthquake when he saw the water in a pond in his garden sloshing. He was worried that his office at Patong Beach might have been damaged by the earthquake. Although it was Sunday, he went to Patong by car, arriving at his office at 9:40. The tsunami attacked shortly thereafter. Thus, people on Phuket Island knew that an earthquake had occurred. However, the tremors of the earthquake were slight, and people received absolutely no information about the location or scale of the earthquake or forecasts about a tsunami during the period of more than two hours between the earthquake and the tsunami attack. As a result, it appears many people went to Patong Beach not knowing the actual situation, and fell victims to the tsunami.

The tsunami attack began when the water started pulling back from the shoreline at around 10:00 a.m. Immediately after this, a rising wave that seems to have been the 1st wave of the tsunami attacked. The depth of inundation in the town facing the coast was about waist deep on people. This sea water quickly receded into the ocean. At this point in time, it seems that people still did not realize that a tsunami was attacking. However, they were at least frightened by the waist-deep water, and everyone began to evacuate from the shore area together. Several minutes later, the larger 2nd wave attacked. In this wave, the water was head-high. Many of the survivors said that they could not have evacuated if the 1st wave had been over their heads, like it was in the 2nd wave.

Regarding the condition of the evacuation, a hotel security guard stated that "People escaped to the 2nd floor, and people who were caught in the current clung to pieces of wood and survived." A certain shop owner said that he escaped to a high building by running and survived. These situations were affected by a distinctive feature of the buildings in Phuket. As can be seen in Fig. 1.21, along the shore, the town consists of a dense cluster of buildings, and these buildings had the effect of a protective wall against the tsunami. Although the front line of wood-frame buildings facing the sea was destroyed (Fig. 1.22), the overall damage to buildings was slight, and the number of damaged or destroyed buildings also decreased with distance from the shore. Damage to buildings was mainly due to inundation (flooding). About 500 m inland from the shore, the inundation depth on roads had decreased to 20 cm. Therefore, the number of buildings which were inundated to the

Fig. 1.21 Densely constructed buildings: Patong Beach.

2nd floor was extremely small. The fact that evacuation was possible in most parts of the town was one factor in the relatively small number of deaths at Phuket (280 deaths) in comparison with Khao Lak (4,200 deaths in Phangnga Province, which includes Khao Lak), as will be described next. However, because the roads were crowded, many people who began to evacuate by car were apparently unable to complete their escape by this means, and many cars were abandoned (Fig. 1.23).

Fig. 1.22 Front line of wood-frame buildings destroyed by the tsunami.

Fig. 1.23 Abandoned cars: Patong Beach on Phuket Island, Thailand.

There was one noteworthy fact in this situation: at Mai Khao Beach on Phuket Island, the wit of a 10-year old British girl saved her parents and many other tourists, and not a single life was lost among them. This will be described in detail later in Chapter 1.4.

The 2nd wave appears to have been also larger at Khao Lak, although some people stated that the 3rd wave was even larger. In any case, there is no question that the 1st wave was smaller. Strangely, many people began to evacuate after the 1st wave at Phuket Island, but on the Khao Lak coast, nobody evacuated after the 1st wave (according to statements by witnesses). Rather, after the 1st wave, many people gathered at the shore, including some who went out to collect fish caught in pools formed on the exposed sea bottom after the wave pulled away from the shore. Many of these people had no time to evacuate when the large 2nd wave hit, and were washed away. What can be guessed from this is that the 1st wave of the tsunami was so small on the Khao Lak coast that people who were on the coast were not frightened, and therefore were late in evacuating.

The situation was the same in hotels. As one feature of the Khao Lak coast, a resort hotel generally consists of wooden cottages scattered around a main building, and the distance from these cottages to the main building is as much as several 100 meters. At around 10:00 on Sunday morning, when the tsunami hit, many tourists were still relaxing inside their cottages. As mentioned above, on the Khao Lak coast, the 1st wave was small, and almost nobody thought that their life was in danger. Because nobody thought anything was wrong, nobody evacuated. In fact, it is possible that people in their cottages did not even notice the small 1st wave. At the sudden attack of the 2nd wave (truly suddenly for those who did not notice the 1st wave), it is supposed that many of the tourists who were in their cottages had no time to escape to the main building, and were swept away by the tsunami (Fig. 1.24).

The Khao Lak coast was attacked by a 10-m high tsunami. Except for strong multi-storey hotel buildings of reinforced concrete construction, all of the buildings collapsed and were washed away. The main buildings of hotels suffered partial damage to windows and walls by the tsunami,

Fig. 1.24 Remains of cottages destroyed by the tsunami (Sunset Beach Hotel).

but the buildings themselves remained intact (Fig. 1.25). Figure 1.26 shows a 3rd floor guest room in a hotel's main building immediately after the disaster. The guests' belongings are scattered about and left, but there are not signs that the room was actually inundated or the walls were damaged. If it had been possible to escape to the main building of a hotel, a large number of people could have taken refuge temporarily in meeting rooms and other large rooms in the hotel and consequently many tourists may have survived in this manner.

People who escaped inland also told what they had experienced. One woman at the Khao Lak coast stated that she escaped by motorbike because the speed of the inundation by the tsunami was too fast to escape by running. The Khao Lak coast is a flat resort area, and there are no embankments or other protective structures on the coast. As a result, inundation by the tsunami spread as far as 1 km inland, making it difficult for people who were late in evacuating to escape to high ground. However, it appears difficult to evacuate by car through chaotic roads. In the fishing port of Baan Nam Khem located north of Khao Lak, elderly grandparents and children who tried to escape by car died. The only

Fig. 1.25 Hotel where windows and walls were destroyed, but the actual structure remained intact.

Fig. 1.26 Hotel guest room (3rd floor).

survivors in the family were the children's parents, who were unable to fit in the car and resolved to remain in their home.

Among unique examples of survival, a working elephant at the coast began to cry an hour before the tsunami attack, and immediately before the attack, the animal escaped and ran to a hill with tourists still on its back. Many people who ran with the elephant were also saved (reported in the Yomiuri Shimbun newspaper, January 4, 2005). At Koh Surin (Surin Island), the Mokaen tribe who live on the island has passed the word from generation to generation that they must flee to the mountains if the sea suddenly pulls back from the shore. Immediately before the tsunami attack, all 600 members of this community fled to the mountains, and the human damage was limited to only three tourists (Mainichi Shimbun, January 11, 2005).

Reference

Hiraishi, T., Arikawa, T., Minami, Y., and Tanaka, M. (2005): Field Survey on Indian-Ocean Earthquake Tsunami-Example Obtained mainly in South-Thailand, Technical Note of the Port and Airport Research Institute, No. 1106, p. 20.

(4) *Disaster in Galle, Sri Lanka (disaster in port town)*

(1) *Outline of Galle*

The city of Galle is located around latitude 6° N on the extreme southern part of the western coast of Sri Lanka. With a population of 104,015 (as of 2001), it is the third largest port city in Sri Lanka (Fig. 1.27). The city extends about 8 km east to west and 7.5 km north to south. The central part of the city covers an area of about 2.5 km east to west and 2.5 km north to south. The central part of the city can be roughly divided into following zones: A fort and the old port are located on land jutting out into the sea on the west side of the city; a main national highway (A2) follows the coastline, and is lined with various commercial buildings as well as public markets; the hilly land to the north is a residential zone; the new Galle Port and the fishery port are on the east side of the central part of the city (Fig. 1.28).

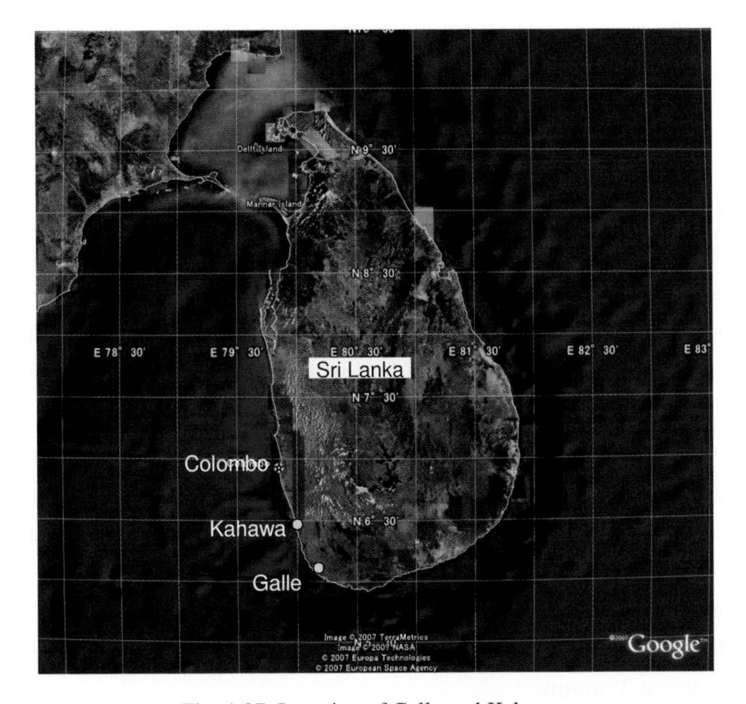

Fig. 1.27 Location of Galle and Kahawa.

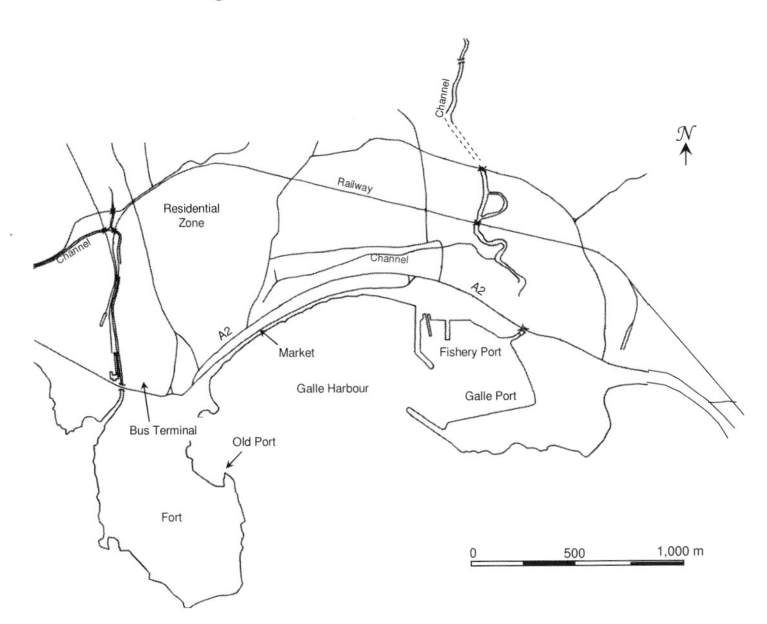

Fig. 1.28 Outline of Galle urban area.

(2) *From the earthquake to the tsunami attack*

Because Galle is located far from the focus of the earthquake, residents did not feel the tremors. The radio news in Sri Lanka had reported the tsunami attack in Indonesia, but those concerned did not associate this with a possible tsunami attack on Sri Lanka because they lacked even a basic knowledge of the nature of tsunamis, namely, that a tsunami in a faraway country can propagate across vast distances of ocean and attack. The memory of past tsunamis in Sri Lanka had only been passed down as a legend. For example, a tsunami had in fact attacked the southern coast of Sri Lanka about 2000 years earlier. However, this was known only as a legend, which described "the sea flooded the land because the Gods were angry with the king's behavior." In 1883, the tsunami that followed the eruption of the Krakatau volcano in the strait between Sumatra and Java of Indonesia, had also attacked Sri Lanka, and a tsunami about 1 m in height had reached Galle. However, the memory of this tsunami had been lost, and was only discovered in written records after the Great Indian Ocean Tsunami.

Almost nobody felt the tremors of the earthquake, and nobody associated the advance information (news from Indonesia) with the danger of a tsunami. As a result, most people fled in order to avoid the flooding induced by the tsunami arriving without warning.

(3) *Tsunami attack and inundation*

The Sumatra Earthquake which caused the Indian Ocean Tsunami occurred on December 26, 2004 at 6:28 a.m., Sri Lanka time (times in the following are Sri Lanka time). The 1st wave of the tsunami attacked Galle approximately three hours later, at around 9:30 a.m., and the tsunami attack continued for several hours thereafter. The city of Galle is located to the west of the southern tip of Sri Lanka, and was thought to be shielded from the tsunami-prone regions by the shape of the land. However, the Indian Ocean Tsunami attacked this location as a result of diffraction by the neighboring coast and reflection from the Indian subcontinent.

Fig. 1.29 Trace of tsunami left on the wall of a building (dotted line).

According to witnesses who saw Galle Bay from a high ground to the west of the city, the backwash current before the tsunami exposed the sea bottom for a considerable distance from the coast of Galle Bay. The tsunami attack continued for several hours after this. Traces of inundation were left on the walls of buildings in the city (Fig. 1.29). Figure 1.30 shows the height (above ground level) of the traces of inundation left on the walls of buildings. The maximum height of traces near the coast and in the port reached approximately 5 m. In the city, the inundation extended about 500 m inland. The height of the traces in high locations was 1-2 m. The tsunami also ran up in irrigation channels and other man-made water channels, and in some places reached more than 1 km inland.

(4) *Inundation of bus terminal*

The Galle bus terminal is located in the center of the city adjacent to the railway station and market, and is bustling with many people on an average day. December 26, 2004 was a Sunday which combined several special factors. Because it was a Full Moon Poya Day (holiday on a day of the full moon) for Buddhists, many believers had closed their shops

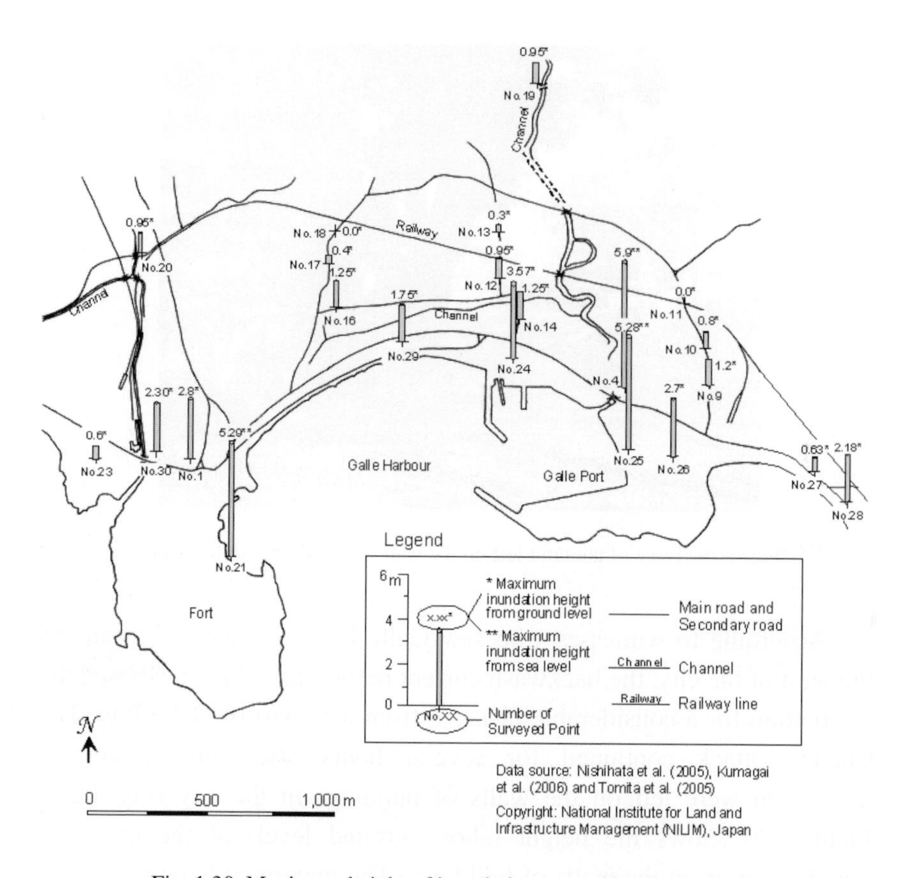

Fig. 1.30 Maximum height of inundation caused by the tsunami.

and were going to temples to pray. It was Boxing Day, which is a traditional day of appreciation for postmen, servants, and others, and it was also the morning of the day after Christmas. For these various reasons, many people were enjoying a holiday from work and were at the bus terminal when the tsunami attacked.

A video of the tsunami shows more than a dozen buses, as well as automobiles, debris, electrical appliances, and other objects floating in the current. People survived by fleeing into buildings facing the bus terminal, climbing onto the roof of the bus stop, climbing trees, and the similar actions. People who were unable to reach such places tried to

Fig. 1.31 Conditions at the bus terminal in Galle during the tsunami attack.

stop themselves from being carried away by clinging to the columns of buildings, but due to the high speed and long duration of the flow, the video shows some who became exhausted and were washed away. The video also shows a person who tried to stand in waist-deep water by holding onto a motorbike. However, due to the fast, waist-high flow (up to about 1 m) and their long duration, it was difficult to withstand the tsunami in this position. Many people climbed onto the roofs of buses, but the buses stalled when the water reached the level of the muffler. The buses stopped running and were floated due to their buoyancy and carried away turning over in the tsunami current. Thus, the top of a bus is not necessarily a safe place (Fig. 1.31).

According to a newspaper report, a man who lost his wife and sons said "My wife and sons were caught in the tsunami and carried far away. Cars near the coast were washed away like leaves from a tree. There were 17 people in one vehicle, but it soon vanished." These examples show that it is not possible to stand in a tsunami current by holding on to some object and cars are easily floated away. It is necessary to flee as quickly as possible from the inundated area. If this is not possible, it is necessary to take refuge on the upper floors (3rd floor or higher) of the nearest building that is safe in the tsunami flow (Fig. 1.32).

Fig. 1.32 Buildings facing the bus terminal (photograph by Kumagai, March 2005).

Fig. 1.33 Bus terminal in Galle immediately after the disaster (photograph courtesy of Mr. H. Gajaba W. Panditha, December 28, 2004).

The bus terminal is located at the neck of a headland-type cape that juts out into the ocean (Fig. 1.28). The tsunami not only reached the bus terminal from the coast facing the direct tsunami attack but also it circled the cape and arrived there from the opposite coast as well (Fig. 1.33).

A monument commemorating the tsunami disaster has been erected in the Galle bus terminal (Fig. 1.34). The inscription reads as follows:

"On December 26, 2004,
The gigantic ocean
Poured in from all directions
And claimed the lives of thousands of our family and friends.
This is a memorial to
All of the victims."

Fig. 1.34 Monument at the Galle bus terminal (photograph by Kazuhiko Honda, February 2007).

(5) *Damage to the port*

In Galle Port, part of the embankment in the port was damaged by the tsunami, and the ground behind it was sucked into the water. A connecting road on the embankment used by traffic in the port was made impassable by this damage. Some cargos which had been stored temporarily in warehouses were also damaged. As mentioned previously, this was the morning of a special Sunday for local people. For this reason, many boats had not gone out that day and were moored in the fishing port. Many of these boats were damaged. The port had a total of 359 fishing boats before the disaster. Of these, 110 were damaged by the tsunami, including 27 which were totally destroyed. An ice-making plant, cold storage facilities, power supply facilities, and other infrastructure were also damaged. Freshly-caught fish will not keep. It can be used for immediate consumption in the area without processing. However, damaged facilities caused problems in supplying to more distant areas and processing.

A crewman on a tugboat in Galle Port spoke as follows about the conditions during the tsunami attack: "The sea surface began to change suddenly. Then, for about 10 minutes, the sea water in the port receded. We hurriedly tried to cast off and get out of port, but couldn't do so because the boat touched the sea bottom and tilted. After that, the sea began to rise again, so the boat wasn't damaged. The sea level rose until it was well over the top of the quay." The water depth in the port is normally 8-9 m. Considering the draft of the boat, the fact that this tugboat touched the sea bottom means that the sea level dropped by several meters. Ships off shore are comparatively safe, but when a ship is tied up in port, like this, it is difficult to get out to sea after a tsunami attack begins. This example speaks of the great efforts of the crew in saving their own lives and their boat while under attack by a tsunami, even though they had no previous experience of this type (Fig. 1.35).

Fig. 1.35 Tugboat that drifted in Galle Port during the tsunami, and its crew (center) (photograph by the author, March 2005).

Fig. 1.36 Ship stranded on land (photograph by the author, March 2005).

In addition, many ships were stranded on land (Fig. 1.36). During a tsunami, it is thought that drifting boats like this one can cause great damage to human life and buildings.

References

Department of Census and Statistics: Population and housing units of DS divisions along the coastal boundaries, 2001.

Domroes, M. (2006): After the Tsunami Relief and Rehabilitation in Sri Lanka … restarting toward the future. Mosaic Books, New Dehli.

Kumagai, K. and Kozawa, K. (2006): Field Survey of Damage due to Indian Ocean Tsunami in Sri Lanka, Technical Note of National Institute for Land and Infrastructure Management, No. 304.

Indo-Asian News Service: Galle is now a city of death and tears, December 28, 2004.

Tomita, T., Honda, K., Sugano, T., and Arikawa, T. (2005): Field Investigation on Damages due to 2004 Indian Ocean Tsunami in Sri Lanka, Maldives and Indonesia with Tsunami Simulation, Technical Note of the Port and Airport Research Institute, No. 1110.

(5) *Train disaster in Kahawa, Sri Lanka*

At Kahawa, a train was swept away by the tsunami, and more than 1,500 people on board died.

The Kahawa District is located approximately 15 km north of Galle. As shown in Fig. 1.37, the area of the disaster is sparsely planted with palm trees. Before they were washed away by the tsunami, the area also included one-storey wood-frame houses and houses constructed of stucco and brick in their scattered state. As shown in Fig. 1.38, the land in this area rises slightly from the coastline, then gently slopes downward and continues as flat land. The rail line is roughly 200 m inland. The height around the rail line is the same as at the coastline, i.e., sea level. When the tsunami attacked, many people were on the train. A typical crowded train is shown in Fig. 1.39.

Fig. 1.37 Aerial photograph of the area around the train disaster (April 22, 2005).

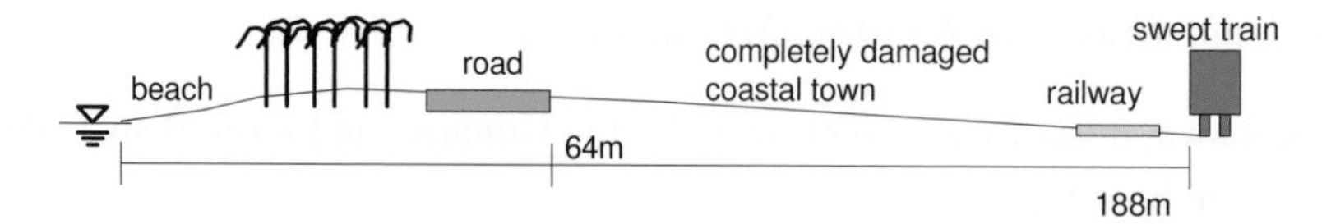

Fig. 1.38 Topography of the land around the train disaster.

Fig. 1.39 Typical train in the Kahawa District (April 20, 2006).

Because Sri Lanka is more than 1,600 km far from the focus of the earthquake, train passengers and residents did not feel the tremors, and because no tsunami information was broadcast, nobody could imagine that a huge wave was about to cause a disaster. According to statements by local people, the tsunami attacked three times. The 1st wave attacked at 9:15 a.m., Sri Lanka time, or approximately two hours after the earthquake in Indonesia. The 2nd wave hit at around 9:45 (according to some statements, around 10:00). The 2nd wave washed away the train and turned it over on its side. The height of the tsunami trace was more than 5 m.

A crew member who worked on the damaged train, Mr. Kalunatekale, stated as follows: "The 1st wave was as high as the floor of the train. It was enough to wet your shoes. However, the 2nd wave was much, much bigger beyond comparison. It bend over the palm trees as it came." From this statement, it can be understood that the inundation height was around 1.2 m in the 1st wave, and was extremely large in the 2nd wave.

The train made an emergency stop because of the 1st wave. A huge volume of water more than 1 m high suddenly came flowing around the train. Because Sri Lanka had not experienced a tsunami attack for more than 100 years, neither the crew nor the passengers understood what was happening when the area around the train was suddenly flooded. Immediately after the disaster, people at the site said they thought "It might have been an atomic bomb test, or a new weapon developed by some country." Neighboring residents also felt something was wrong. There were many wood-frame and brick houses in the area, and some people thought the stopped train would be a safer shelter than their own houses. As a result, the residents were divided into those who fled inland and those who took shelter in the train. While they were doing so, the 2nd wave struck.

The height of the 2nd wave can be estimated from the trace left on a reinforced concrete building nearby. This building was located closer to the coast than the rail line, and was about 200 m away from the place where the train was overturned. Tsunami trace was found on the wall of the building at a spot 4.2 m above the ground surface (Fig. 1.40, left). Because this wall was on the back side of the building from the direction of the tsunami attack, the maximum tsunami height was probably higher than this. The position of the tsunami trace have now been marked on this building (Fig. 1.40, right).

Fig. 1.40 Building near the rail line (left: January 6, 2004, right: March 12, 2006).

The 2nd wave of the tsunami was about as high as the train cars. When struck by this tsunami wave, the cars of the train, including the engine, were swept away and scattered in different directions. One railway car was carried almost 70 m away (Fig. 1.41). Figure 1.42 shows the site immediately after the disaster. The tsunami attacked from the right side in the photo. The tracks were twisted up off the ground, and the train was hurled into the neighboring palm grove. This clearly shows the power of the tsunami. More than 90% of the people who took shelter in the train died in this tsunami.

One reason for the enormous power of this tsunami was, of course, the size of the tsunami itself. However, the shape of the land was also a factor. The disaster area lies along the coast, and as mentioned previously (Fig. 1.38), the land rises from the shore to a coastal road, and then slopes downward toward the railway tracks. The land is also flat for several kilometers in the landward direction, and includes ponds here and there.

Because the area was at the coast, the tsunami formed a bore-type wave and broke. As a result, the wave approached at a faster speed

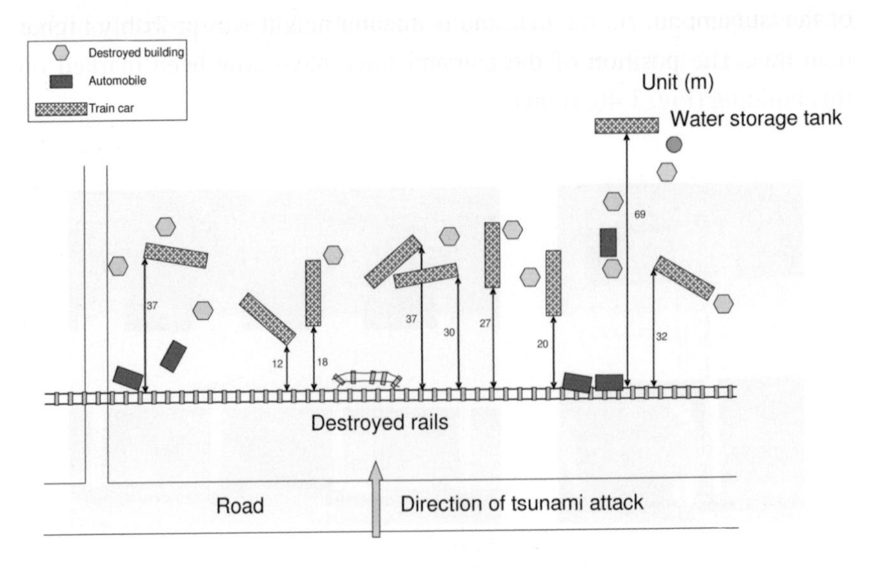

Fig. 1.41 Train cars scattered by the tsunami.

Fig. 1.42 Site immediately after the disaster (©AP Images).

than normal. It also accelerated as it traveled down the slope before hitting the train. In technical terms, it had become a "supercritical flow." As can be understood from Fig. 1.37, the area contained palm trees and scattered houses, but these were not a significant obstacle to this powerful tsunami. As a result, the tsunami struck the train with its force undiminished, and swept the train away with all the people on it.

There were three basic causes for the increased human damage in this tsunami. There was a time lag of about 2 and 1/2 hours between the earthquake occurrence and the tsunami attack. Furthermore, because this area is located in the southwestern part of Sri Lanka, there was a time lag of about 15 minutes with the eastern part, which was damaged first. The fact that this time lag could not be used effectively, for example, to transmit information (warnings), was the first cause that worsened the disaster. After the 1st wave of the tsunami attack, many passengers and residents took refuge in the train. Because local residents took refuge, they obviously expected something bad to happen after the 1st wave. However, they did not know that the 2nd and 3rd waves would be even larger than the first, even though this frequently occurs in tsunamis. Thinking they would be safe on the train, they decided to use it as a place of refuge. This was the second cause that worsened the disaster. However, even assuming these first and second causes had not existed (that is, adequate warning had been given and people had avoided the train), there was still a third cause, which was the fact that the land is flat for a long distance inland, and there is no high land or strong tall buildings near the coast.

Finally, let us cite the testimony of one survivor, Mr. Silver, who was on the train with his parents and three children, but miraculously survived the disaster. As the water flooded the train compartment, he took only one of his children, little Tahiru, on his back and tried to escape through a window of the overturned train. As he climbed out, he pushed away people who were clinging to his legs. His parents and his two other children died. He continued to say, "It was a terrifying wave. The water came just like a bulldozer pushing up the ground. The wave was pitch black. The people I was pushing away might have been my own parents, but at least I saved this one child."

References

NHK Special (February 27, 2005): "The Great Indian Ocean Tsunami – The Reality in Images."

Tomita, T., Arikawa, T., Yasuda, M., Imamura, F., and Kawata, Y. (2005): Field Survey around South West Coast of Sri Lanka of the December 26, 2004 Earthquake Tsunami Disaster of Indian Ocean, Annual Journal of Coastal Engineering, Japan Society of Civil Engineers (JSCE), Vol. 52, pp. 1406-1410. (in Japanese)

1.3 Hokkaido Nansei-oki Earthquake Tsunami Disaster (Japan, 1993)

(1) *Outline of the tsunami disaster*

(1) *The earthquake and tsunami*

The Hokkaido Nansei-oki Earthquake struck at 10:17 p.m. on July 12, 1993 at north latitude 42°47' and east longitude 139°12' (Fig. 1.43). The depth of the earthquake was 34 km and the magnitude was 7.8. Large tremors with seismic intensities on the Japanese scale of 5 to 6 were felt over a wide area centering on Japan's large northern island, Hokkaido. The tsunami hit Okushiri Island, located about 60 km southwest of the seismic focus, five minutes after the earthquake, and caused massive damage.

The focus of this earthquake lay on the interface between the Eurasian Plate and the North American Plate. In the past, earthquakes occurred in this area roughly every dozen years, but only 10 years before the Hokkaido Nansei-oki Earthquake, a large earthquake (M = 7.7) struck at 11:59 a.m. on May 26, 1983, at north latitude 40°21' and east longitude 139°02' (Fig. 1.43) and triggered a tsunami that reached Okushiri 17 minutes later.

(2) *Outline of the region*

Okushiri Island is an isolated island located in the Sea of Japan approximately 20 km southwest of Hokkaido. The island is roughly 10 km wide east to west and 20 km long north to south, and has a coastline of approximately 84 km. Its area is 149 km^2 and the population is about 4,000 persons, most of whom are engaged in fishing for a living. The fishermen's homes were near the coast for convenience of fishing, while most governmental and commercial buildings were also located in the plains near the coast.

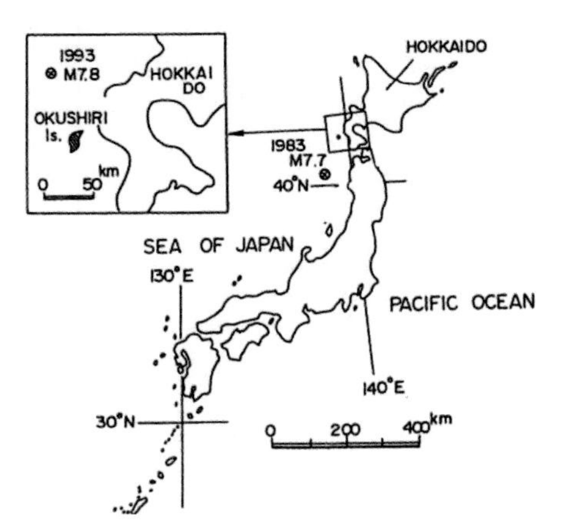

Fig. 1.43 Location of focus of the Hokkaido Nansei-oki Earthquake.

(3) *Features of the tsunami*

The seafloor around Okushiri Island has a steep slope of 1/5 to 1/2, and so as the tsunami approached the island its height increased rapidly. Figure 1.44 shows the tsunami runup heights at various parts of the island. (Note: "Runup" means the climb of the water on land measured from sea level. "Inundation" means the water depth above the ground surface.) The tsunami encircled the entire island, and the runup was especially high at the ends of the headlands that jut out into the sea. In particular, the tsunami concentrated at the shallows located at Cape Aonae at the southern tip of the island, causing the runup height to increase there. Villages in the low-lying areas near the coast suffered catastrophic damage upon being hit directly by the tsunami.

Because the tsunami occurred at night (10:22 p.m.), many older persons were already asleep in bed. Although there are almost no video records of the tsunami attack, witnesses reported that they heard loud rumbling sounds, which were immediately followed by the tsunami.

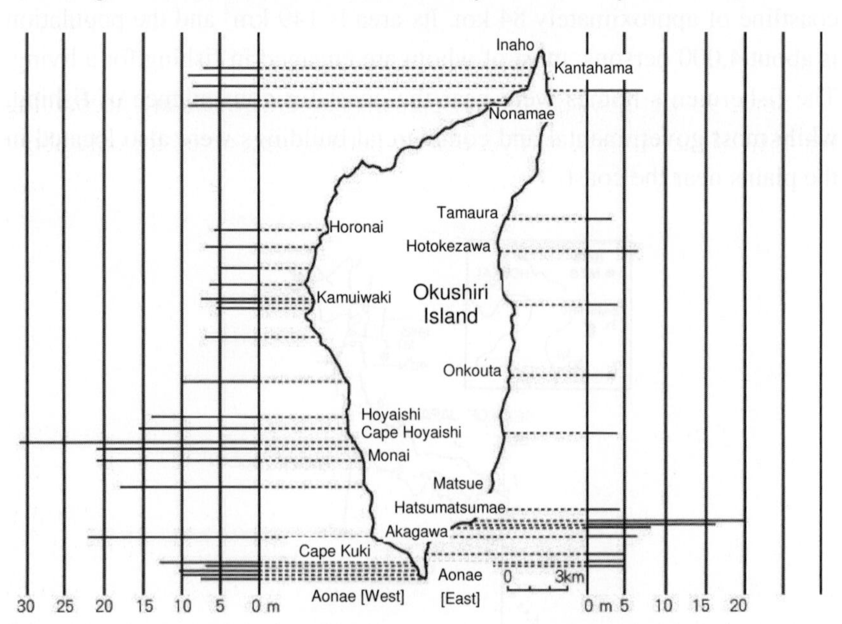

Fig. 1.44 Runup height of the tsunami around Okushiri Island (showing the Aonae area at the bottom of the map).

(4) *Outline of damage*

After the tsunami caused by the Nihonkai-Chubu Earthquake in 1983, a 4.5-m high concrete seawall was built at Okushiri Island as a "hard" countermeasure. However, the tsunami in 1993 was far larger than expected and the seawall was almost completely ineffective. An evacuation advisory system had also been constructed as a "soft" countermeasure. In 1993, an evacuation advisory was issued on the judgment of the person in charge of disaster prevention in the Okushiri Town Office. The widespread experience that a tsunami follows an earthquake resulted in an extremely high evacuation rate of 70%. However, because the tsunami struck just five minutes after the earthquake, there was very little time to evacuate and many were killed while evacuating.

The death toll on Okushiri Island due to the earthquake and tsunami was 198, or about 5% of the island's population. The overwhelming majority of deaths were caused by drowning (138 persons), followed by death due to crushing (18), brain contusions (15), fire-related injuries (2), and other causes (29).

Some 90 victims were elderly persons (aged 61 or older), accounting for 45% of the victims, and yet elderly persons accounted for only about 15% of the island's population. Thus, the death rate of the elderly was relatively high.

(2) *Tsunami disaster in Aonae District of Okushiri Island*

(1) *Outline of Aonae District*

Aonae District at the southern tip of Okushiri Island is divided into seven wards (1st–7th Wards). The 1st to 5th Wards along the coast are on flat ground just 2–8 m above sea level. Houses were concentrated around the Aonae Fishing Port. Public facilities such as fishery and agricultural cooperatives, clinics, post offices, junior high schools, etc., as well as inns and restaurants, were also located in the same area. The 6th and 7th Wards were located on higher ground, 30 m above sea level, and

Fig. 1.45 Appearance of Aonae District on the morning after the tsunami (July 13, 1993).

included public housing, the Aonae Branch of the Okushiri Town Office, the police station, and the fire station.

The population of Aonae District at the time of the earthquake was 504 households, totaling 1,401 persons. Human loss in the tsunami was 87 dead and 20 missing. Physical damage included 342 houses either totally or partially destroyed. The most severe damage was in the 5th Ward, which suffered 67 dead, 5 missing, 6 seriously injured, and 8 with minor injuries. At the time of the disaster, the 5th Ward had 79 households, totaling 213 people. Thus, about 1/3 of the ward's residents died. Many of the houses which were not washed away by the tsunami were destroyed by fire. One day after the earthquake, the area resembled a burned-out plain (Fig. 1.45).

(2) *From the earthquake to the tsunami*

Mr. Kiyoshi Orito, who was in charge of disaster prevention at the time of the earthquake, has the following memories of the conditions just after the earthquake: "I was at home, relaxing before bed, at 10:17 p.m. when the earthquake hit. Suddenly I felt a vertical tremor. I knew right away that this wasn't an ordinary earthquake. I was first reminded of the Nihonkai Chubu-oki Earthquake 10 years before. I waited for the tremors to stop, and then went to the Town Office. I got there at about 10:20, and I broadcast a warning, 'Danger of tsunami, evacuate!' to the whole island through the wireless disaster prevention system." It has not been confirmed how well the system in the Okushiri Town Office actually communicated this message to the island's residents. However, it is reported that the evacuation order was successfully issued to the residents before the first tsunami struck at 10:23 p.m.

The Japan Meteorological Agency issued its tsunami warming at 10:23, based on which the national television network, NHK, broadcast an emergency warning at 10:25. However, the first tsunami wave had already hit Okushiri Island.

(3) *Tsunami attack and inundation*

Figure 1.46 shows an outline of the tsunami attack on Aonae District. The first tsunami wave arrived from the west about five minutes after the earthquake and was 7–9 m high. The tsunami passed over the 4.5-m high levee on the west side, struck the village, and then destroyed the concrete seawall on the east side. The second wave hit from the east side of the cape 12 to 15 minutes later. This tsunami, which was 10 m high, passed over the green belt and destroyed a large number of houses. The third and fourth waves passed through the fishing port and destroyed facilities of the fishing port and houses in the hinterland.

Mrs. Miyuki Ono, who lived in the 5th Ward and was rescued after being washed away with her house by the tsunami, reported as follows: "When I ran out of the house clutching the children and started up the road leading to the hill, I could see something white in the pitch dark. That was the tsunami. Immediately after that, the wall of the next house collapsed onto us and I was sandwiched in the debris with the children. I shouted for help and somebody noticed, but he had to run away because of the second wave of the tsunami, but luckily it didn't reach us. The

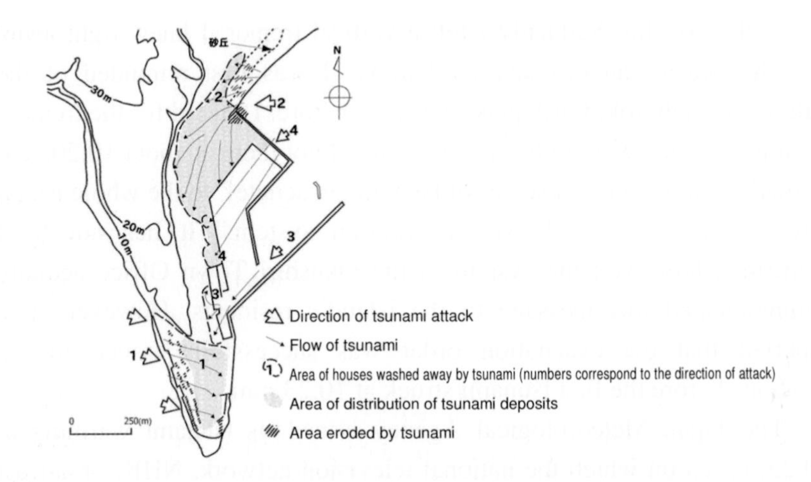

Fig. 1.46 Tsunami attack at Aonae (Tsutsumi et al., 1994).

confusion in the area continued in the dark for a while. About three hours later, we were finally rescued. We were transferred to Sapporo by helicopter and survived." Mrs. Ono survived the tsunami because she found a tiny space in which to shelter between the crushed buildings and because the second tsunami wave did not reach her.

A large number of people were rescued while drifting at sea after being washed away to the ocean by the tsunami. One of them, Mrs. Toshiko Karihara, spoke as follows: "The tsunami came with a sound like sand being swept up in a river. The approaching tsunami knocked me off my feet and I fell down. I was carried out to sea by the backflow. After being carried several kilometers offshore, I noticed a half-destroyed boat and caught it. I floated on the tide with the boat along the coast and I was rescued by a fishing boat four or five hours later. I was able to drift for such a long time because it was mid-summer and the water was warm, and also because the ocean was dead calm with no waves. Since I am confident about swimming, I didn't panic. I optimistically thought that things would get better if I just went on floating, but if it had been the Sea of Japan in mid-winter, I would not have been so optimistic — the water in winter is so cold that you would die in just 10 minutes." As these comments suggest, the death toll would have been higher if the tsunami had struck in winter.

(4) *Evacuation from the tsunami*

In Aonae District, there was a traffic jam caused by people trying to escape to higher grounds in their cars, and so many people left their cars and tried to evacuate on foot, but were caught by the tsunami. Evacuation by car might seem wise, but it can actually take more time if you get caught in a traffic jam. This example confirms that the rule is to evacuate on foot in case of a disaster, except for elderly and physically handicapped people.

In this tsunami, many people survived because they evacuated quickly wearing only the clothes on their backs. In contrast, those who died (1) could not run, (2) waited until all family members were together, (3) hesitated to leave family who could not evacuate by themselves, or (4) stopped to warn neighbors. Many other victims were young children or older people who straggled behind during the evacuation.

Looking at the distribution of the number of dead by house in the 5th Ward of Aonae District (Fig. 1.47), it is remarkable to note that many people who lived far from high land were able to escape but many people who lived near there could not. It seems that these people felt safe and delayed starting to evacuate because they lived near the high land of refuge areas.

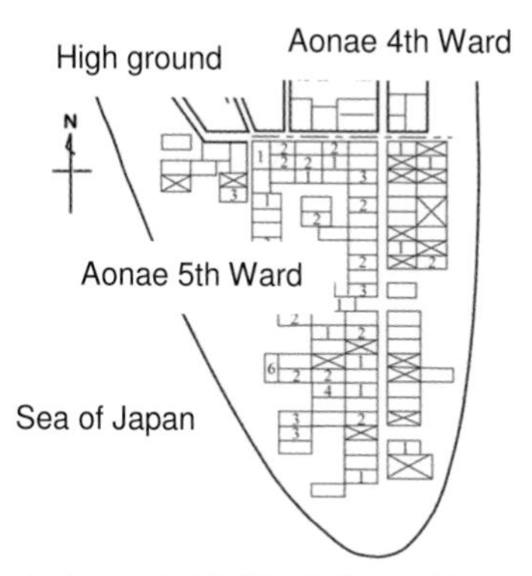

No. of deaths by house in 5th Ward, Aonae District, Okushiri-cho

Fig. 1.47 Deaths by house in the 5th Ward of Aonae District (Shuto, 1993).

(5) *Damage to buildings*

Around 10:30 p.m. on July 12 just after the tsunami attack, two fishing boats moored in Aonae Fishing Port caught fire, and then the fire spread to nearby houses. At 12:15 a.m. on July 13, a different fire broke out at another place, and home-use propane gas cylinders and kerosene tanks ignited and fed the fire. Because many of the residents had evacuated and were not at home, firefighting could not be performed adequately at first. In addition, the debris carried by the tsunami blocked the roads and fire trucks could not reach the fire, thus causing worse damage.

As a result, Aonae District burned to the ground. The fire was finally put out at 9:20 the next morning (July 13). This was a major fire disaster with a total of 189 buildings burned, including 108 households totaling 311 persons. The total floor space burned was over 50,000 m^2.

(6) *Experience of fishermen*

Mr. Takami Matsumae, who was fishing on the east side of Okushiri Island when the earthquake occurred, stated that he "heard a knocking sound on the bottom of the boat, but I didn't really think it was a tsunami because the boat hull didn't shake very much." However, he added that "I had a wireless device on the fishing boat at the time. If I had been able to feel a tsunami while I was offshore, I would have warned people on the beach to flee."

We should also mention the fishermen who died in the tsunami in areas other than Okushiri Island. At Setana Town in Hokkaido, four of the six victims were fishermen, swept away by the tsunami as they pulled up their small boat at a slipway near their homes. Their bodies were discovered in the sea nearby. In Kamoenai Village, a fisherman and his daughter were swept out to sea while driving along a coastal road to check on a fishing boat some distance from their home. Several hours later, the car was caught in a fishing net with its headlights still on, but the two persons in the car were already dead. These victims died while trying to protect their fishing boats. They had enough time to finish the work to protect their fishing boats after the tsunami ten years before. This experience might have caused the fisherman's poor judgment.

1.4 Case Study — What Separated Life and Death?

(1) *Process until safe evacuation*

Some amount of time passes between an earthquake (or other event causing a tsunami) and a tsunami attack. Therefore, if people sense the danger in advance and evacuate to a safe high place or building, they can escape the tsunami. This means that survival is determined by whether people can receive and understand various information immediately after an earthquake, judge the danger, and based on this, evacuate quickly and correctly.

Looking at past examples, there are three stages in safe survival, as follows:

(1) Information is necessary in order to begin an evacuation. Typical examples of this information are earthquake tremors and tsunami warnings. First, these must be received and understood.
(2) In some cases, a single piece of information is not enough to begin an evacuation. People often do not think evacuation is necessary due to some kind of wrong preconception. Judgment of evacuation is made by a decision-making process using various pieces of information.
(3) Simply beginning an evacuation does not ensure survival. Danger can only be avoided by moving from your present location (home, etc.) to a safe place by various evacuation routes. The important point is this: You must be aware of danger along the evacuation route. In some cases, however, depending on the timing of the tsunami attack, it is more important than anything else to escape quickly to some high place.

In each process from (1) to (3), various judgments separate life and death.

(2) *Information about danger — How can I recognize danger?*

A public tsunami warning is a typical example of information announcing the danger of a tsunami. In addition to information about the earthquake (location, magnitude), this information also includes the predicted arrival time and height of the tsunami. Japan's Meteorological Agency now has a system that issues tsunami forecasts within two minutes after an earthquake. This is the world's leading tsunami warning system, not only in content, but also in speed and reliability. Tsunami information is not limited to broadcast using the public disaster prevention wireless system. It is also transmitted to local areas and individuals by television and radio, the internet (public websites), and other channels. In the Sumatra Earthquake/Indian Ocean Tsunami of 2004, there was no tsunami warning system of this type. This was one factor that caused such a large number of deaths. Furthermore, the tremors could not be felt except Indonesia and others which were located near the center of the earthquake. A tsunami warning would have been the only advance information of danger if any had been there. Because there was no tsunami warning system in the region at the time, coastlines around the Indian Ocean were struck by a massive tsunami without any advance warning.

Earthquake tremors, strange changes in the ocean as well as abnormal sound and wind and other signs on the coast also work as advance information on tsunamis. However, it is difficult for an individual to judge whether a tsunami will attack or not. For example, earthquake tremors are affected by the ground and other conditions, it is not possible to judge whether the earthquake occurred on land or at sea, and whether the earthquake was large enough to cause a tsunami, from the ground motion alone. The judgment about the earthquake ground motion in the Showa Sanriku Earthquake Tsunami is an interesting example of this problem, as will be discussed in detail in Case Study 1 below. In this case, a particular preconception passed down through generations in the region or held by individuals delayed the evacuation. Another sign of a tsunami is unusual changes in the ocean. However, by the time these are recognized, the tsunami is already within sight, and immediate judgment is necessary. In the Indian Ocean Tsunami, some people recognized a tsunami attack was imminent even without a tsunami warning, and saved those around them. Case Studies 2 and 3 discuss the examples of a young British girl in Thailand and a Buddhist in Sri Lanka who saved lives in the Indian Ocean Tsunami.

Case Study 1: The Meiji Tsunami and Showa Tsunami in Sanriku — Evacuations delayed by wrong preconception

On June 15, 1896 (Meiji 29), the sky over the Sanriku coast was covered with heavy clouds typical of the rainy season in Japan. Rain had been falling off and on since morning, and the weather was extremely hot and humid. An earthquake occurred a little after 7:30 in the evening. The tremors were not strong, and the intensity on the coast was around 3. This earthquake was later called the "Slow Earthquake." However, the tsunami that followed was enormous with a maximum run-up height well over 38 m and caused catastrophic damage on the coastline. Because the tsunami was so huge and caused such great damage, it left a vivid memory in many people's minds. Unfortunately, the combination of conditions on this day led the local people to various ideas that were hardly scientific. For example, "when tremors are weak, a large tsunami will occur" and "large tsunamis occur in summer during humid, and rainy weather." As these ideas were passed down, they gradually changed to the beliefs that "when tremors are strong, the tsunami will be small," "tsunamis don't occur in winter," and "tsunamis don't occur on clear days."

Then, 37 years later, another earthquake occurred on the Sanriku coast, on March 3, 1933 (Showa 8). This time, the conditions were completely different: The earthquake was powerful with an intensity exceeding 5, and it occurred in winter, on a cold clear night when the stars were visible. According to the beliefs handed down in the area, a tsunami could not occur under these conditions. As a result, there was no alert, and many people did not evacuate. But a large tsunami did come.

The residents of the Hongo area of Touni Village, Kesen-gun were awakened by a powerful earthquake and were worried about a tsunami, so they grabbed their important belongings and hurried to high ground with their families. At the time, older people were confident that "a tsunami won't come in clear weather." Because it was bone-chilling cold outside, most people wanted to believe the old people. These people returned to their homes and went back to bed, and it was too late when a tsunami attacked them. There was a similar incident in Taneichi Village. In a certain family, a younger sister who had lived through the Great Kanto Earthquake (Tokyo, 1923; more than 100,000 deaths) had come back to the village and was staying there. She was also confident that there would be no tsunami, saying "There's nothing to worry about. The tremors were far stronger in the earthquake in Tokyo, and they were different, vertical movements. This time the tremors are horizontal." She said, "If you don't want to catch a cold in this freezing weather, you all should go back to bed." They were also late in evacuating. (This account is from Akira Yoshimura, "Great Tsunamis on the Sanriku Coast," Bunshun Bunko; in Japanese, p. 191.)

Case Study 2: A young British girl in Thailand — Lives saved by education in school

The Indian Ocean Tsunami caused a catastrophic disaster on the Andaman Coast, which includes Thailand's Phuket Island resort area. The dead totaled more than 8,000 and included foreign tourists on year-end holidays. No tremors were felt here, and many lives were lost when the tsunami attacked without warning. Only Maikhao Beach was an exception, with no deaths or missing persons. This miracle was thanks to a 10-year-old British girl named Tilly. The Sun newspaper, which presented the story as a scoop, named her the "angel of the beach."

Tilly and her family had come from Britain to Phuket Island on vacation. On the day of the tsunami, the family was all enjoying the sun of this southern country on Maikhao Beach. Looking out to sea, they realized that something was strange. The ocean seemed to be foaming, and then suddenly the tide pulled back from the beach and the water level fell. Tilly realized immediately that this meant something. She had learned about earthquakes and tsunamis at school. What's more, her teacher had explained that the sea suddenly pulls back about 10 minutes before a tsunami attacks. "I knew right away that something was going to happen. I was sure it was a tsunami, so I told my mother."

In many cases, an adult would ignore the child, but fortunately, Tilly's parents didn't doubt their daughter's intuition. Other people quickly responded to the warning by Tilly's family. Everyone on the beach, and everyone in the hotels facing the beach, quickly moved inland before the terrifying tsunami hit. Thanks to Tilly, several hundred people escaped the tsunami.
http://www.thesun.co.uk/article/0,,2-2004610510,00.html

Case Study 3: The Buddhist of Sri Lanka — Remembering an old story

This is the experience of a young girl in the Balapitya District, north of the city of Galle in Sri Lanka. When the tsunami attacked, she was still sleeping at home, but she realized something was wrong when the 1st wave struck the coast without a sound. When she looked at the coast, it was in a strange condition as the sea level had risen about 1.5 m. Sensing danger, she tried to wake her younger brothers, who were sound asleep after a party the day before. Because she couldn't wake them, she pushed them out of the bed. Next, the water pulled back. Fish and other things could be seen on the dry sea bottom. This was not normal. She saw fishermen at sea. After that, the large 2nd wave hit. It was black and had a rotten smell. She had sensed danger when the 1st wave struck because she immediately remembered a story about a tsunami in Buddhism.

Western Sri Lanka was ruled more than 2,000 years ago by King Devanampiyatissa (in the reign: 247-207 B.C.). One day, a deity was enraged because the king had killed a Buddhist monk, and made the sea engulf the land. In this legend, the sea flooded the land 15 miles (24 km) from the coast and swallowed up many people. To soothe the deity's

Fig. 1.48 The princess, who is being sacrificed, boards the boat of gold (wall painting in the Raja Maha Vihara temple).

anger, the king made a boat of gold and set his oldest daughter adrift in it as a sacrifice (Fig. 1.48). When he did, the sea retreated. The boat eventually landed at Kirinda, on the southern coast of Sri Lanka. The lucky princess who survived this terrible ordeal married King Kavantissa, who united this region at the time, and was named Queen Viharamahadevi. A monument commemorating these events can be found in the outskirts of Kirinda.

(3) *Understanding and judgment of information*

Many people will not act on only one piece of information. For example, few people will move to higher ground simply because they felt earthquake tremors. Some people may even go down to the coast to check for signs of a tsunami attack. Without a belief that danger exists, which is formed by obtaining and arranging multiple kinds of information, people will not evacuate. In Japan, the Meteorological Agency monitors the evacuation rate that shows the percentage of people who evacuate when a tsunami warning is issued. This rate has been falling in recent years. Individuals will not take action if they are not given information that exceeds a level where their risk recognition is triggered. When knowledge and experience are poor, their risk recognition abilities are also poor and extremely large amount of information is necessary. Actually, on the Sanriku coast and elsewhere, there are many examples in which people only evacuated when their neighbors called out even though they felt the tremors of an earthquake earlier. Case Study 4 discusses the Showa Sanriku Tsunami.

Case Study 4: In Taro Village, Iwate Prefecture in the Showa Sanriku Earthquake Tsunami — Neighbors' voices press residents to run away

This is the same tsunami as in Case Study 1. However, the actions of the residents were completely different. On March 3, 1933 (Showa 8), the residents of Taro Village were awakened by an earthquake accompanied by violent rumbling of the earth. Houses shook noisily, and objects on shelves fell. People felt anxious. The memory of the Great Tsunami 37 years before was still fresh, and the older people said that a tsunami might attack this time as well. Some of the careful men went outside to check for signs of a tsunami. Because they didn't notice any of the abnormalities usually seen before a tsunami, they were reassured. Everybody went home and went back to bed.

However, the quiet village suddenly heard the warning blast of a steamship's horn from the ocean. In some houses, people jumped out of bed at the sound. Words expressed with some doubt "Is tsunami coming?" were changed to more definitive "Tsunami is coming!," which then quickly spread in the neighborhood. In an instant, the village became a scene of noisy confusion. People flew from their houses and ran into the darkness, fleeing to the mountain behind the village. Some were pushed and fell, and soon the narrow road was crowded with people, children clinging to any adults around them and sick people crawling on the ground. People too weak to walk were just sitting there.

All of a sudden, hearing the definitive information "Tsunami is coming!", every villager began evacuating simultaneously, creating a panic. (From Akira Yoshimura, "Great Tsunamis on the Sanriku Coast," Bunshun Bunko; in Japanese, p. 191.)

People judge whether to evacuate or not after receiving information about a disaster (earthquake tremors, tsunami warning, etc.). In doing so, they use different judgment standards, depending on their individual knowledge and experience. The example of the evacuation during the Hokkaido Nansei-oki Earthquake in 1993 is a good lesson. First, in 1983, many deaths occurred as a result of the Nihonkai Chubu Earthquake Tsunami off northern Akita Prefecture. On Okushiri Island, which is located to the southwest off Hokkaido, the tsunami attack occurred about 20 minutes after the earthquake, and the coastal area was damaged. Only 10 years later, in 1993, a giant earthquake occurred just to the west of Okushiri Island. This time, the tsunami arrived three minutes after the earthquake, swallowing the island. Residents who evacuated in this very short time survived. However, people who believed their own past experience thought they still had time and did not evacuate immediately. It was the moment that separated life and death. Case Study 5 describes the events on Okushiri Island.

Case Study 5: Two Tsunamis on Okushiri — The Nihonkai Chubu Tsunami of 1983 and Hokkaido Nansei-oki Tsunami of 1993

On May 26, 1983, an earthquake centered off northern Akita Prefecture caused a tsunami that reached the coast of the Sea of Japan. The tsunami hit Okushiri 15-20 minutes after the earthquake, killing two persons. One was a visiting sport fisherman, and the other, an islander (working fisherman) living in the Aonae District. This islander delayed his evacuation to inspect his boat on the beach after the earthquake.

Ten years later, after 10:00 on the evening of July 12, 1993, another earthquake occurred near Okushiri Island. At the time, Aonae 5th Ward was home to 77 households, totaling 214 people. Only 3-5 minutes after the earthquake, a tsunami more than 10 m high struck the west side of the island. At the time of this earthquake, the Meteorological Agency

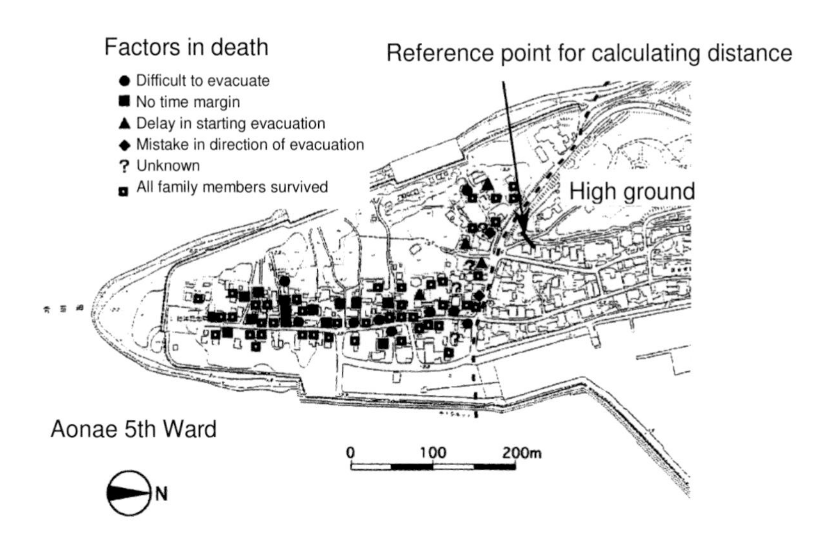

Fig. 1.49 Factors in death by household in Aonae 5th Ward (In this figure, the 5th Ward is located to the left of the dotted line).

issued a "large tsunami warning" five minutes after the quake. Although the households on the island had installed disaster prevention wireless after the tsunami 10 years earlier, this because the tsunami attack occurred before the tsunami warning was issued. Virtually every house in the 5th Ward was totally destroyed, and 72 persons died or went missing. Almost everyone had experienced the tsunami 10 years earlier, but in some households all family members survived, and in others, some did not. Looking at the geographical distribution of the survivors, there was no relationship between the survival and the distance from the coast or the distance from high ground (Fig. 1.49).

The statements of survivors include comments like "I had just gotten out of the bath, so I threw a towel over my shoulders and ran," "I escaped in my pajamas," "I flew out of the house before the shaking stopped and escaped in my car," "I ran like a sprinter, heading for the high ground." Almost everybody who survived ran outside as soon as the earthquake stopped and escaped by running. These people began to evacuate immediately, irrespective of the fact that the tsunami attack 10 year before hit 20 minutes after the earthquake.

How did the victims react? It seems that they either did not evacuate or could not evacuate, or they were late in evacuating. Some persons were caught in the tsunami because they ran in the wrong direction (toward the tsunami) or tried to evacuate using roads along the coast. Among those who were late in evacuating, witnesses reported people who were walking slowly, changing clothes, looking for car keys in the dark, strayed from the route to a grandmother's house in an effort to evacuate together, or stopped to warn relatives living nearby. According to a statement by one survivor, he went outside, saying "A tsunami may be coming, let's escape by car." One person said "Just a minute, I want to make a call," and began telephoning. The survivor said, "Hurry up, let's go!" and got in his car, which was stopped in front of a shop. He waited for the other to finish his call, and when he looked back, the second floor of the house had collapsed. He saw a white wave higher than the roof. He quickly closed the door, and the next instant, the car was pushed sideways and rolled over into the ocean, where it was swallowed up by the tsunami. One of these persons survived, and two others died. The victims may have learned nothing from the tsunami 10 years earlier.
(Source: Asahi Shimbun, "That Night in Okushiri," Okushiri reporting team, Asahi Shimbun, 1994, p. 190.)

(4) *Judgment while evacuating*

It is important to evacuate immediately to a safe place, but this may not ensure survival. In many cases, people are caught by the tsunami while evacuating, depending on the route and means of movement. The key point is to move from your present location (home, etc.) to a safe area by appropriate routes. Remember that it may not be possible to use roads where you normally travel by car. Roads may be destroyed by the earthquake, crowded by mass evacuation from the tsunami, or blocked by debris by the tsunami. Looking at the geographical distribution of the victims in Aonae District on Okushiri Island (Fig. 1.49), the survival rate does not necessarily increase closer to high ground. In addition to delays in evacuation, the reasons for this include mistakes in the direction of evacuation and in the selection of the evacuation route. Case Study 6 also discusses the tsunami at Okushiri.

Case Study 6: Examples from Okushiri — Choices that determined survival

"In many cases people who evacuated on foot but were swept away by the tsunami were saved while floating at sea. Almost everybody in cars died." This is from the statement by Mr. Takayoshi Kimura of the Okushiri Town Office, who participated in the recovery work. In fact, Mr. Hayashi, who had put out to sea on a squid fishing boat, rescued around 10 people from the waves. If the arrival time of the tsunami had been 20 minutes after the earthquake, as it was in the 1983 tsunami, evacuation by automobile might have been the correct choice. However, in the 1993 tsunami, the 1st wave stranded a number of cars on the road, and the gigantic 2nd wave hit where the road was blocked by these cars.

Many residents escaped to high ground and let the tsunami pass. Again, Mr. Kimura reported that "many of the islanders were aware of the need to escape the tsunami." What saved Mr. Arakawa and others was quick evacuation. However, Mr. Arakawa commented that "there was no comparison between the previous tsunami and the one in 1993." In the 1993 tsunami, the maximum height was 30 m, and in some places, the tsunami hit only 2-3 minutes after the earthquake. The tsunami also changed direction depending on the topography of the coastal seafloor and the shape of the bay, and there were repeated attacks. Mr. Seiji Kagaya, who lost his mother in the tsunami and later headed the Okushiri Survivors Association as its chairman, survived the 1st wave by fleeing to the mountain behind the village, but he was caught in the 2nd wave when he tried to return home. "I didn't think there could be another one bigger than the 1st," he said. "And the wave came from the direction of the mountain. It was unimaginable." He survived by clinging to a tree.
(Source: Chunichi Shimbun, November 1, 2002, "Tsunami" (Be Prepared: Part 7))
http://www.chunichi.co.jp/saigai/jisin/feature/2002110101.html

Reference

Motoyuki Ushiyama, Motoko Kanata, and Fumihiko Imamura: A Study on Mitigation in Human Damage by Disaster Information for Tsunami Disaster, Journal of Natural Disaster Science, Vol. 23-3, pp. 433-442, 2004.

1.5 Damage to Buildings

(1) *Classification of damage to buildings*

Needless to say, the degree of damage to buildings depends on the size of the force acting on the building. It is possible to estimate the flow velocity and force (fluid force) of a tsunami from the height of the tsunami and the depth of inundation. Therefore, in classification of the degree of damage to buildings on land, the depth of inundation is frequently used as a guideline for external force, because a comparatively large amount of data is available on this factor. (For structures located inshore, the height of the tsunami is generally used as a standard.) However, problems remain, as the flow velocity is not necessarily the same with a given inundation depth. For example, as the distance from the shore increases, the flow velocity of the tsunami, and its fluid force, generally decreases even when the inundation depth is still large. If a cliff is located directly behind a building, the degree of damage also generally decreases even when the inundation depth is large.

The degree of damage to buildings also depends on the construction and structure of the building, the distance from the sea, the topography behind the building, and other factors. Types of construction include wood, stone, brick, block, reinforced concrete, and others. Structural factors include the arrangement and weight of the walls and columns, methods of joining columns and beams, the construction of walls, etc. Moreover, the degree of damage may not be uniform in buildings of the same construction, because the quality of the materials used, structure, and strength requirements, also differ depending on the country and region. For example, in reinforced concrete (RC) buildings, the strength of buildings will differ greatly in a building with reinforced concrete walls and a building with unreinforced brick walls.

Therefore, the following will discuss the degree of damage depending on differences in the construction of structures by inundation depth. All of this knowledge was gradually clarified by repeated site surveys after tsunami disasters.

In damage to buildings, impact by floating objects can be a major factor. For example, after the Hokkaido Nansei-oki Earthquake Tsunami, a building which had been half-destroyed with a maximum inundation depth of only 0.25 m was discovered. Floating objects during the tsunami were considered to be the main cause of this damage. Sometimes, the impact force of floating objects is greater than the simple fluid force of the water.

When it is necessary to judge the degree of damage to a building based on whether it is reusable or not, damage caused by factors other than physical forces could also be a problem, which includes the damage to buildings by oil and other flammable substances, toxic substances, dirty mud, and the like. Figure 1.50 shows the condition in front of the Great Mosque in Banda Aceh in Northern Sumatra after the Indian Ocean Tsunami in 2004. Although the water depth was only 1.6 m here, mud and debris made the building unusable.

Fig. 1.50 Building damaged by sludge and debris (Matsutomi et al., 2006).

(2) *Classification and examples of building damage by inundation depth*

When the inundation depth is less than 2 m, damage basically does not reach "heavy damage" whether they are wood, stone, brick, block, or RC buildings. In a small fishing port, a tsunami with a height of 3.7 m and inundation depth of 2 m struck a house located near the quay, but because the structure was anchored to a concrete foundation, damage was limited to some columns and walls (Fig. 1.51). This house was repaired and is being reused.

When the inundation depth is 2-3 m, in many cases, wood-frame buildings will suffer heavy damage, while stone, brick, and block buildings will suffer light to moderate damage, and RC buildings will suffer light damage only.

Figure 1.52 is an example of a wood-frame house part of which suffered heavy damage by a tsunami with a height of 6.7 m and inundation depth of 2.1 m. A road runs in front of this house and on the opposite side of the road, there used to be a group of wooden warehouses starting from around the central position in the figure and

Fig. 1.51 Damage to a wood-frame house by a tsunami with inundation depth of less than 2 m (Yunoshiri fishing port, Akita Prefecture, Japan).

Fig. 1.52 Wood-frame house part of which suffered heavy damage and is badly leaning due to an inundation depth of 2.1 m.

Fig. 1.53 House that was protected by a group of warehouses in front of it.

extending toward the right. (These warehouses suffered heavy damage and the remains have already been removed.) Beyond these former sites of warehouses, there is a seawall, and then the sea. In this figure, the tsunami struck from the left front. When evacuation is not executed properly, even a tsunami attack of this size will cause deaths, and in fact, there was one death in this area. Figure 1.53 shows the damage to a house located only 30-70 m to the right of the area in Fig. 1.52. This house was protected by the group of wooden warehouses standing over the entire area on its coast side. (These warehouses suffered heavy damage; in the figure, their remains have already been removed and the site has been graded.) The warehouses reduced the inundation depth to 1.4-1.8 m, and as a result, the damage to the house was only moderate. This is an example of how structures in front of a building can protect the building in the rear.

Figure 1.54 shows the damage to a brick building in a tsunami with a height of 5-6 m and an inundation depth of approximately 2 m. The wall is of brick-and-stucco construction. (The wall itself is brick and mortar, with stucco finish on the front and back sides). The wall was unreinforced. This type of wall is easily destroyed by inundation on the order of 2 m.

In Fig. 1.55, the height of the tsunami around the group of houses in the center of the photo was 4.0-5.1 m, and the inundation depth was 2.3-2.9 m. The past record shows that this inundation depth generally causes heavy damage to wood-frame houses. Here, however, the percentage of heavy damage was as low as around 50%. If buildings are densely clustered, and structures in front of the buildings protect those behind, some buildings can withstand the action of the tsunami.

When the inundation depth is 3-7 m, in many cases, wood-frame buildings suffer heavy damage; stone, brick, and block buildings suffer moderate damage; and RC buildings receive light damage.

Figure 1.56 is an example of the damage to a building of RC construction. At the building on the left edge of the figure, the tsunami height was 10.0 m and the inundation depth was 5.5 m. Although this building was inundated to the 3rd floor, it suffered only moderate damage, which was very close to light damage. In this case, people survived by taking refuge on the 3rd floor. On the left, the eaves of the building's roof were destroyed. This is thought to be due to the mass of water which shot up when the flood current hit the building. In the case

Fig. 1.54 Brick house the wall of which was destroyed by inundation of 2 m (Phuket Island, Thailand).

Fig. 1.55 Example of clustered buildings that withstood a tsunami (Kainan Town, Japan, 1986).

of strong buildings such as reinforced concrete structures, if damage to the main structure can be avoided, the building can also protect those behind it from the tsunami. From this viewpoint, constructing public buildings in reinforced concrete on the coast side, and constructing individual houses behind these structures, is a way to reduce tsunami damage. Individuals can also reduce tsunami damage by constructing homes with an RC garage as the ground floor.

In Fig. 1.57, damage was reduced by structures in front of the buildings. The inundation depth was 4-6 m. In a group of buildings located nearer the sea (in the foreground of the figure), only their concrete floors remain. The buildings at the rear are more or less intact.

Even when damage is only moderate, it may become impossible to reuse the building. For example, in Fig. 1.58, the building in the center is leaning because the ground under it was eroded by the tsunami current. Damage due to ground erosion must be considered when discussing tsunami refuge buildings and other facilities.

When the inundation depth reaches 7 m, even stone, brick, and block buildings also suffer heavy damage. However, the probability of heavy damage in RC buildings is unknown, as there have only been two cases to date where RC buildings were subjected to inundation of 7 m or more.

Fig. 1.56 Reinforced concrete buildings that received moderate damage with an inundation depth of 5.5 m (Khao Lak, Thailand, 2004).

Fig. 1.57 Example in which damage was reduced by structures in front of them (Matsutomi et al., 2006).

Fig. 1.58 Building leaning due to erosion of the ground (Matsutomi et al., 2006).

Figure 1.59 shows the damage to a reinforced concrete house located 0.9 km from the coast at Banda Aceh, Northern Sumatra, after the 2004 Indian Ocean Tsunami. Although the structure is reinforced concrete, the walls are of unreinforced brick construction. Because the water was blocked and rose at the front of the building, the inundation depth was around 8 m. Damage was judged to be "moderate," but was close to "heavy." The reason why damage did not reach "heavy" may be because the walls were easily destroyed, and this reduced the fluid force on the structure. This house was still in this condition about 20 months after the tsunami. Whether the family that lived here survived or not is unknown. However, they may have survived if they took refuge on the 2nd floor believing the structure itself would withstand the tsunami.

Figure 1.60 shows the houses destroyed and washed away by the 10 m tsunami in the Sanriku area of Japan on March 3, 1933. The figure at the left shows Taro Village before the tsunami; that at the right shows

Fig. 1.59 House in which some RC columns and the walls were destroyed by inundation of 8 m.

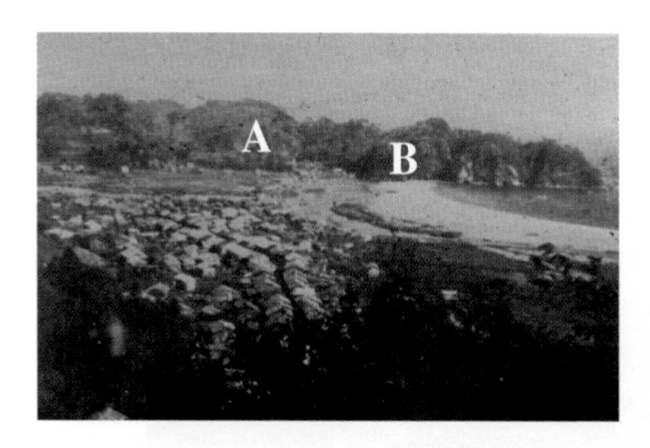

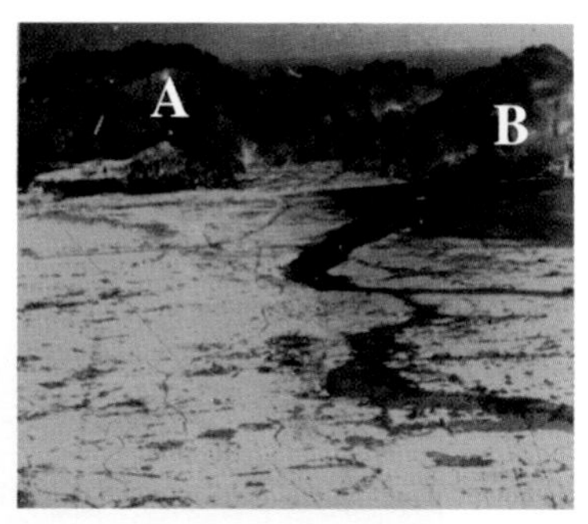

Fig. 1.60 Wooden houses washed away by the Showa Sanriku Tsunami (Taro Village, Iwate Prefecture, Japan; left: before the tsunami (1932) and right: after the tsunami (morning of March 3, 1933)).

the same area on the morning after the tsunami struck around 2:00 in the morning. The white area in the photo at the right is snow. Points A and B in the figures are the same points (the figure at the right was taken from the same direction, but is a closer view). From these before-and-after figures, it is clear that virtually every house in the village was destroyed and washed way. All of these buildings were wooden houses. In this district, 358 out of 362 houses were destroyed and washed away, and 763 out of 1,798 persons either died or went missing.

(3) *Damage to buildings by floating objects*

Buildings may also be damaged by floating objects. In the past, typical examples of floating objects included fishing boats and other small vessels, and driftwood. More recently, floating objects also include shipping containers, automobiles, and oil tanks. Floating oil and other flammable substances and harmful chemicals cannot be ignored.

Examples of buildings destroyed by drifting boats are too numerous to list them all. Figures 1.61 and 1.62 show examples from Kuwagasaki Town (now Miyako City), Iwate Prefecture, Japan after the Meiji Sanriku Tsunami of 1896 and Ofunato City, also Iwate Pref., after the 1960 Chile Earthquake Tsunami. A large sailboat of the time and a 250-ton fishing boat crushed wooden houses.

Fig. 1.61 Destruction of wooden houses by drifting boats (courtesy of Iwate University, Iwate Pref., 2006).

Fig. 1.62 Example of destruction of a wooden house by a drifting boat (Ofunato City, 1962).

Fig. 1.63 Power generating barge weighing 2,500 tons, which was carried approximately 3 km inland in the Great Indian Ocean Tsunami (Banda Aceh, Northern Sumatra, Indonesia, 2004).

Among larger vessels, in particular, flat-bottomed boats with shallow drafts can be carried far inland, causing massive damage. Figure 1.63 shows a representative example from the 2004 Indian Ocean Tsunami. Here, a power generating barge weighing 2,500 tons, which had been moored at Banda Aceh in Northern Sumatra, Indonesia, was carried approximately 3 km inland from the port, causing great damage to buildings.

Driftwood always causes some degree of damage to buildings when a tsunami floods inland. In the 1960 Chilean Earthquake Tsunami which reached Japan, driftwood from a lumber mill in Shizugawa Town, Miyagi Pref., Japan destroyed wooden houses. The destructive forces acting on buildings may be increased even further if fishing boats, driftwood, and other floating objects are caught across two buildings, creating a large surface area blocking the flow and in turn increasing the inundation depth. In addition to the force of the tsunami itself, these forces can also cause destruction or washing away of buildings.

References

Iwate Prefecture, Iwate University (2006): Tsunami Disaster Prevention Training Materials: – Protecting Children from Tsunamis – , DVD, No. 1. (in Japanese)

Ofunato City (1962): Chilean Earthquake Tsunami of 1960, Ofunato Disaster Journal, p. 415. (in Japanese)

Kainan Town (1986): Record of the Nankai Earthquake Tsunami, "Fated Asakawa Port," p. 116. (in Japanese)

Fire Defense Agency (1965): Research on the Niigata Earthquake Fire, p. 224. (in Japanese)

Matsutomi, H. (1990): "Wave Force due to Breaking Wave/Bore Wave Collision Driftwood – Case of Similar Bore Height and Driftwood Diameter – , Proceedings of the Coastal Engineering Committee, Japan Society of Civil Engineers (JSCE), Vol. 37, pp. 654-658. (in Japanese)

Matsutomi, H. and Shuto, N. (1994): Tsunami Inundation Depth and Flow Velocity and House Damage: Proceedings of the Coastal Engineering Committee, JSCE, Vol. 41, pp. 246-250. (in Japanese)

Matsutomi, H. and Iizuka, H. (1998): Tsunami On-Land Flow Velocity and Its Simple Estimation Method: Proceedings of the Coastal Engineering Committee, JSCE, Vol. 45, pp. 361-365. (in Japanese)

Matsutomi, H., Suzuki, A., Kurizuka, K., and Sato, K. (2005): Actual Condition of Embankment Breakage Flooding and House Destruction due to Niigata/Fukushima Heavy Rains in July 2004: – Kariyatagawa Nakanoshima District, Igarashigawa Suwa/Mabuchi District – , Tohoku Journal of Natural Disaster Science, Vol. 41, pp. 75-80. (in Japanese)

Yamashita, F. (1982): "The Tragic Story of the Sanriku Great Tsunami," Seijisya, p. 413. (in Japanese)

Matsutomi, H., Sakakiyama, T., Nugroho, S., and Matsuyama, M. (2006): Aspects of Inundated Flow due to the 2004 Indian Ocean Tsunami, Coastal Engineering Journal (CEJ), JSCE Vol. 48, No. 2, pp. 167-195.

Shuto, N. (1993): Tsunami Intensity and Disaster, Tsunamis in the World, Kluwer Academic Publishers, Dordrecht, pp. 197-216.

1.6 Damage by Ships and Other Floating Objects

(1) *Outline of damage*

Ships, which are anchored in ports, are normally moored using mooring ropes/cables and fenders. However, depending on the size of the tsunami, mooring ropes may come loose or break, allowing ships to drift freely. Because ships are designed to float on water, a loose ship will simply "go

Table 1.4 Relationship between size of tsunamis and damage caused by ships and other floating objects.

Size of tsunami	Ship/floating object	Mode of damage
A: Relatively small tsunami (height: 2-3 m or more)	Small boats	Drifting Collision with quays, etc. Capsizing/sinking Stranding on land
	Lumber	Drifting Left on land
	Shipping containers	Drifting
	Automobiles	Carried into sea
B: Relatively large tsunami (height: 5-6 m or more)	Small boats	Stranding on land Collision with buildings, etc.
	Larger ships	Drifting Collision with quays, etc. Stranding on land Collision with buildings
	Automobiles	Drifting Collision with buildings, etc.

with the flow." Similarly, because wood floats on water, lumber is easily tossed about by tsunami currents. Some ports have timber basins where timber is stored floating in seawater. If a tsunami hits one of these timber basins, the wood may be carried away. In addition, empty shipping containers and automobiles are frequently carried away by tsunamis, becoming dangerous floating objects.

Table 1.4 shows the relationship between the size of a tsunami and the damage caused by ships and other floating objects. In general, the degree of damage increases as the size of the tsunami increases and the size of the ship or other floating object decreases. In damage caused by ships in past tsunamis, fishing boats and other small craft frequently caused damage. Damage by drifting ships occurs when the height of the tsunami reaches around 2 m. When the height of the tsunami is 2-3 m, small boats may be carried away as floating objects, capsized and sunk, crash into quays, or left stranded on top of quays and jetties. Lumber stored in timber basins and empty shipping containers have also been carried away by tsunamis of this height, and automobiles and other objects have been carried into the sea by the return flow of the tsunami. When the height of the tsunami reaches 5-6 m, not only small boats but also larger ships are carried away and left stranded on the land. These ships can cause serious human injury and damage to buildings. Care is necessary, because small boats such as fishing boats, lumber, shipping containers, and other objects can be carried away by even a small tsunami. As the height of the tsunami increases, larger ships, automobiles, and other large objects are also carried away and drift. The damage caused by the tsunami will increase even more if these floating objects collide with buildings or people.

(2) *Drifting of ships*

In fishing ports, many fishing boats are moored at quays or pulled up on slipways. Because small boats like fishing boats are easily affected by a

Fig. 1.64 Fishing boats stranded on a beach (courtesy of the Port and Airport Research Institute (PARI), Japan, Reference 1).

tsunami, they will drift freely if their mooring ropes come loose or break. Furthermore, fishing boats that are pulled up on land are generally left where they can easily put out to sea again. Therefore, if a tsunami is large enough, these are easily carried away. For example, even in a 1.5-m tsunami, fishing boats that had been pulled up on shore have been carried away and drifted, interfering with other fishing boats that were trying to evacuate. Thus, care is necessary, as small boats in a port may drift even in a small tsunami. Ships and boats are carried onto land by tsunamis when the height of the tsunami is larger than the height of the land above sea level. Normally, the height of piers in fishing ports is at most about 2 m. If the height of a tsunami exceeds this, fishing boats will be carried onto land. In some examples, fishing boats were moved to and fro on piers by the push and pull of a tsunami when the tsunami height was around 3.5 m. Figure 1.64 shows fishing boats which drifted up and were stranded on land. The height of the tsunami was around 5 m. The boats in the photograph had been moored in the fishing port, but were carried out of the port by the tsunami, drifted, and eventually were carried up on this beach and left stranded. Larger ships will not drift unless the height of the tsunami is 5-6 m or more. This is because large ships are not pulled up on land, but rather are firmly moored to a quay. As a result, larger ships are less easily affected by a tsunami than small boats.

(3) *Capsizing, sinking, and collision of ships*

Ships which are carried away by a tsunami may be tossed about and capsized by violent currents, or may fill with sea water and sink. If only some of a boat's mooring ropes break or come loose, the moored boat may become unbalanced and capsize. Even when none of boat's mooring ropes were cut, fishing boats have been capsized by the return flow of a tsunami. This capsizing and sinking of boats occur even in small tsunamis. In a certain 2-m tsunami, fishing boats that had been moored to the quay were capsized and the bottoms of some boats were damaged by contact with the sea bottom during the return flow. Figure 1.65 shows a capsized fishing boat in a port. The height of the tsunami was around

Fig. 1.65 Capsized fishing boat (PARI, Reference 2).

Fig. 1.66 Collision involving two fishing boats (PARI, Reference 2).

2.5 m. This port had a breakwater constructed of wave-dissipating blocks, but the tsunami destroyed the breakwater, and a nearby fishing boat was capsized. Figure 1.66 shows fishing boats that have collided beside a quay. The tsunami height was 3.5 m or more. Here, collision mode involves a fishing boat riding up on top of another boat. In this tsunami, many fishing boats were lifted and left stranded on the quay and breakwater, or were capsized and sunk. When the height of the tsunami reaches around 5 m, fishing boats drift in the port, while being tossed about by the violent currents. Some sink, and others are carried up onto land. In a tsunami of this size, a cement transport ship weighing approximately 7,000 tons capsized and a container ship weighing approximately 10,000 tons crashed into the breakwater while entering the port.

(4) *Boats lifted onto land and collision with buildings, etc.*

There are many cases in which boats that have been carried onto land and drifted in a tsunami are left on land when the tsunami retreats. Figure 1.67 shows a fishing boat that was lifted by a tsunami and ran up on a quay. Here, inundation by the tsunami with a height of approximately 2.5 m reached over the top of the quay, and the drifting fishing boat was left stranded on the quay. As shown here, fishing boats can be run up on breakwaters and be left stranded on quays by tsunamis with a height of 2-3 m. Figure 1.68 shows a cargo ship carried onto a pier. Here, the cargo ship weighing several hundred tons was run up on the pier by a tsunami with a height of around 5 m. The same tsunami carried many fishing boats and cargo ships onto land. These vessels were left stranded on embankments and roads, in fields and rice paddies, etc. when the tsunami retreated. When the height of the tsunami was the same 5 m, a 1,000-ton work ship moored in a port was run up on a quay. Because many box-shaped ships, such as work ships, have a shallow draft, the

Fig. 1.67 Fishing boat that ran up on quay (PARI, Reference 3).

Fig. 1.68 Damage involving a cargo ship carried onto a pier (Suzaki Port pamphlet, Reference 4).

Fig. 1.69 Fishing boat that collided with a house (PARI, Reference 5).

possibility of being carried onto land by a tsunami is high. When the tsunami height reaches around 7 m, even large cargo ships and other similar vessels may be carried onto land. In one case, a power generating barge (approx. 2,500 tons) which was moored in a port was carried by a tsunami to a residential area 3 km inland.

When a tsunami is relatively large, ships carried onto land by the tsunami can collide with and destroy houses and other buildings, causing enormous damage. For example, there are cases where a 4 m-scale tsunami struck a fishing port, ran up on a boat slipway area or passed over an embankment and the boats were carried away, colliding with houses. In a 5 m-scale tsunami, a cargo ship weighing several hundred tons can be carried away and strike houses. In cases like these, a large number of drifting fishing boats destroy houses, and even cargo ships moored in the port are carried onto land, where they can crush buildings, increasing the damage. Figure 1.69 shows a fishing boat which was carried onto land and the damage it caused to a house.

(5) *Evacuating and securing the safety of ships and boats*

It is generally thought that ships and boats are safer on the sea outside a port than inside the port. This is because the height of a tsunami is not particularly large on the open sea but increases rapidly in ports. It has been reported that ships on the open sea received no damage during a tsunami which recorded a height of 5 m in the port. There is a case where after receiving a warning over phone that a tsunami was approaching, the majority of about 200 fishing boats in the port began to evacuate to sea and escaped damage. Therefore, when an earthquake occurs and a tsunami attack is predicted, ships and boats in ports should evacuate immediately to the sea outside the port, if time allows. In this case, it is important to catch tsunami forecast information as quickly as possible. In one example, a tsunami with a height exceeding 4 m attacked a short time after an earthquake. A total of approximately 200 large ships and fishing boats were moored in a port. However, the damage was concentrated on the smaller boats. The larger ships were equipped with telecommunications equipment and received early warning of the tsunami. Approximately 100 of these larger ships evacuated outside of the port and escaped damage. Other large ships could not evacuate, but were able to take emergency countermeasures, such as increasing the number of mooring cables. These also avoided damage. On the other hand, the smaller vessels with poor communications equipment did not receive adequate warning. If ships in ports can evacuate to sea promptly after a tsunami warning is issued, it is possible to avoid damage to the ships themselves. In addition, human injury and damage to houses and other buildings by drifting ships can also be reduced. However, care is necessary in the timing of evacuations of ships. In another example, fishing boats were trying to evacuate out of a port when the tsunami attack was at its peak, but they were pushed back at the port entrance and could not go any further. When the expected time until the tsunami attack is too short for ships to evacuate, if time allows, it is preferable to take countermeasures such as increasing the number of mooring cables and ropes. If there is no time for emergency countermeasures, people must evacuate to safety immediately. As mentioned previously, boat owners have died in past tsunamis because they went to the port to check on their boats.

(6) *Damage by collision with drifting lumber and containers*

Some ports have timber basins, which are used to store wood materials by floating in seawater. In some cases, lumber is stored on land near lumber mills. Lumber stored in a timber basin can be carried away by a small tsunami with a height of less than 2 m. In a certain example, a 2-m tsunami carried away approximately 15,000 pieces of lumber in a timber basin. Of this, about 700 pieces were carried out of the port. Figure 1.70 shows drifting timber after a 5-m tsunami. Large pieces of lumber piled at a lumber mill near the coast were carried away. This drifting lumber injured people, and destroyed houses, allowing them to be swept away. As this example shows, lumber is easily carried away by a tsunami, and become dangerous drifting objects that can cause serious injury.

Fig. 1.70 Drifting lumber (source: "Warning from the Sea," Reference 6).

Fig. 1.71 Drifting containers (PARI, Reference 3).

In some ports, shipping containers are stored in container yards. Care is necessary, as empty containers are easily floated by a quite small inundation depth. Figure 1.71 shows drifting containers after a tsunami approximately 2.5 m in height. The containers were carried out into the port. Drifting containers as these also pose a serious danger to people and buildings.

(7) *Other drifting objects*

Other drifting objects include aquaculture rafts, furniture, remains of houses, automobiles, train cars, oil tanks, and other large objects. Aquaculture rafts, furniture, and debris from houses are as dangerous as timber and can harm people and destroy buildings. Wooden houses which are not secured to a foundation can easily float and drift. As for automobiles, there is a case where automobiles parked in port parking lots were floated and carried into the sea by tsunamis no more than 1.5 m in height. In another case, the return flow from a 3.5-m tsunami swept about 50 cars in a ferry parking lot from the quay into the sea. Figure 1.72 shows the scene after an attacking tsunami passed over an embankment. The return flow from this tsunami swept cars parked on the embankment into the sea. As seen in these examples, automobiles parked near the coast can be swept into the sea by a tsunami.

Fig. 1.72 Drifting automobile (source: "History of Disasters in Japan", Reference 7).

Railway trains and other large objects are not necessarily safe. In the Great Indian Ocean Tsunami of 2004, coastal residents took shelter in a train that was stopped by inundation by a 5-m tsunami. In this terrible tragedy, more than 1,000 persons died when the next wave hit the train (described in 1.2(5)). In a 7-m tsunami, which struck Northern Sumatra in the same Indian Ocean Tsunami, oil tanks (height: 11 m) drifted approximately 300 m along the coastline, eventually reaching a town. The three drifting tanks were all empty. As these cases show, large structures are easily carried away by tsunamis when their weight is small.

References

Tomita, T., Honda, K., Sugano, T., and Arikawa, T.: "Report on Site Surveys and Numerical Analysis of Damage in Sri Lanka, the Maldives, and Indonesia by the Indian Ocean Tsunami," report of the Port and Airport Research Institute (PARI), No. 1110, p. 36 (2005). (in Japanese)

Tanimoto, K., Takayama, T., Murakami, K., Murata, S., Tsurutani, K., Takahashi, S., Morikawa, M., Yoshimoto, Y., Nakano, S., and Hiraishi, T.: "Actual Condition and Second and Third Observation of 1983 Nihonkai Chubu Earthquake Tsunami," PARI report No. 470, p. 299 (1983). (in Japanese)

Tomita, T., Kawai, H., and Kakinuma, T.: Damage Caused by the Tokachi-oki Earthquake Tsunami and Features of the Tsunami, PARI report No. 1082, p. 30 (2004). (in Japanese)

Ministry of Land, Transport and Infrastructure, Shikoku Regional Development Bureau, Kochi Port and Airport Construction Office; Suzaki Port pamphlet. (in Japanese)

Takayama, T., Suzuki, Y., Tsurutani, K., Takahashi, S., Goto, T., Nagai, N., Hashimoto, N., Nagao, T., Hosoyamada, T., Shimosako, K., Endo, K., and Asai, T.: Features and Damage of the 1993 Hokkaido Nansei-oki Earthquake Tsunami, PARI report No. 775, p. 225 (1994). (in Japanese)

Suzaki City, Kochi Prefecture: Record of the South Pacific/Chile Earthquake Tsunami, "Warning from the Sea," Suzaki City, Kochi Pref., p. 151 (1995). (in Japanese)

Shimozuru, D., Tsumura, K., and Miyazawa, S.: History of Disasters in Japan, Vol. 2, Earthquakes and Tsunamis, Nihontosho Center Co., Ltd., p. 207 (2001). (in Japanese)

Chapter 2

Knowledge for Tsunami Survival

2.1 Tsunamis and the Coast

(1) *Tsunamis in offshore areas (water depth >30 m)*

Here, "offshore" means the open sea extending beyond the horizon where the water depth is more than 30 m. The tsunami source area, where tsunamis are generated, is generally in this area. Because a person standing on the shore cannot see these waters, it is difficult for persons to sense an approaching tsunami directly.

The "tsunami source area" is the area of the ocean where a tsunami is generated. A tsunami is generated when the sea bottom ground is displaced vertically by an earthquake, and this displacement is transmitted to the sea surface, where it causes a similar displacement. As will be discussed later, the sea surface is displaced very slowly at this time, and the swelling of the sea surface can hardly be felt on ships navigating these waters. However, it has frequently been reported that crew members on ships sailing in a tsunami source area felt a trembling short-period vibration during the earthquake. This vibration is called a "seaquake." This is thought to be a type of shaking due to elastic vibration of the water caused by the short-period component of the sea bottom ground movement. To date, there have been no reports of ships that were damaged by seaquakes. A small boat was capsized in Toyama Bay (Japan) during the Noto Earthquake on March 25, 2007. It has been pointed out that its cause was a seaquake, but the facts in this case are unclear.

Horizontal deformation of the sea bottom ground does not cause large changes at the sea surface. However, large vertical displacement of the sea bottom ground causes a corresponding large motion at the sea surface. In the tsunami source area, the sea bottom ground may move a few meters (at maximum, about 5 m) in the vertical direction, and it is estimated that this motion is completed in the duration of earthquake tremors (1-2 minutes). From this, the deformation rate of the sea bottom ground is small, being no more than 5 cm/s. Thus, it is an extremely gradual deformation. If the large vertical displacement progresses at a rate of no more than 5 cm/s and the deformation takes place over 1-2 minutes, the deformation of the sea surface may take a different shape from the vertical deformation of the sea bottom ground. However, because the size of the tsunami source area is so large ranging from several 10 km to several 100 km, the sea surface is displaced in basically the same shape as the displacement at the sea bottom ground, except at the edges of the deformation region. Because the displacement takes place over such a large area, even if there is some deviation in the deformation at the edges, this difference induced by this displacement is very small, and its effect on the tsunami is also small.

Figure 2.1 shows the vertical displacement of the sea bottom ground by the 1498 Meio Earthquake (M=8.6), which occurred off the Pacific coast of Japan. The solid line shows areas where the ground rose (upheaval), and the dotted line shows areas that sank (subsidence). It is estimated that the sea bottom ground rose approximately 4 m in this earthquake. Based on the fact that the subsidence occurred near land, while the upheaval occurred on the open sea side, the ground was moved by the earthquake in such a way that it rode up from the land side toward the open sea side, forming a reverse fault. Because the sea bottom sank on the land side, it can be estimated that the tsunami which reached the coast began with a backwash (receding water). According to seismologists, when the fault plane is parallel to land, reverse faults probably occur because the land side is heavier, and tsunamis tend to begin with a backwash. There is a traditional belief that vigilance against a tsunami is necessary when the sea water suddenly pulls back far out to sea. This belief is correct; in many cases, tsunamis begin with this kind of backwash. However, it is not correct to say that tsunamis always begin with a backwash. If the sea bottom on the land side rises, the tsunami

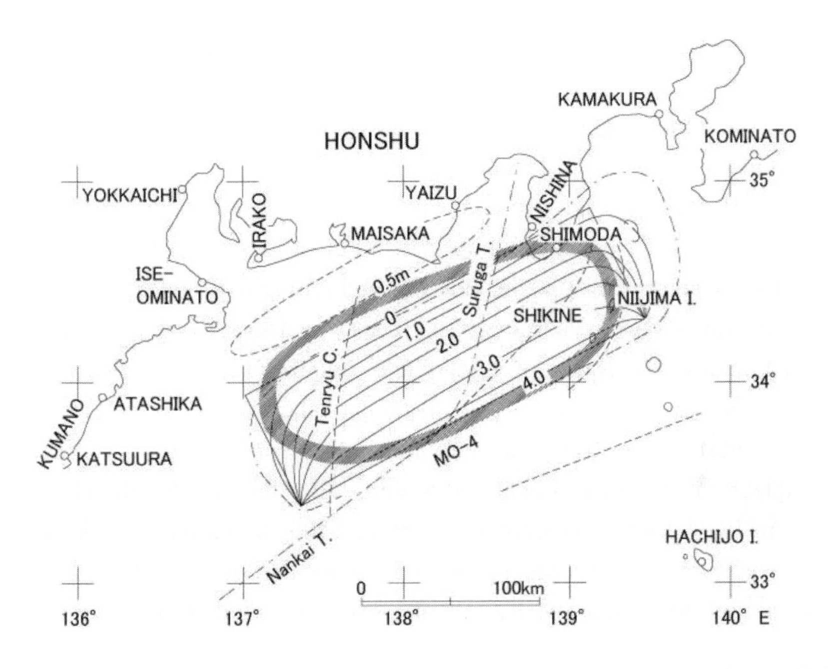

Fig. 2.1 Vertical displacement of the sea bottom by the Meio Earthquake (M=8.6, September 20, 1498).

will begin with sea water surging onto land. There is a slight, but very important difference in the expressions "When the sea water suddenly pulls back to sea, a tsunami will attack" and "When a tsunami attacks, the sea water will suddenly pull back to sea." The first is true, but the second is wrong.

A tsunami is a phenomenon in which the deformation of the sea surface caused by the displacement of the sea bottom ground propagates, radiating in all directions as a wave motion due to the action of gravity. Because the tsunami source area is large, the period of tsunamis is long, at more than 10 minutes. Therefore, in water depths from shallow (<10 m) to very deep (>10,000 m), the speed at which a tsunami radiates, or tsunami propagation velocity, C is given by the following equation:

$$C = \sqrt{gh} \tag{1}$$

where, h is the water depth (m) and g is the acceleration of gravity (9.8 m/s^2).

In the middle of the Pacific Ocean, where the average water depth is 4,000 m, tsunamis propagate at approximately 720 km/h. This is roughly the same speed as a jet airplane. The tsunami caused by the *M*9.5 giant earthquake off the coast of Chile in 1960 crossed the Pacific Ocean and reached Japan on the opposite side in roughly one day, and caused heavy damage on Japan's Pacific coast. The waveform of the tsunami is transmitted at the speed of a jet plane, but the water particles do not actually move at this speed. Transmission of the tsunami waveform (that is, the shape of the water surface) and movement of water particles are completely different physical phenomena. In other words, although the tsunami waveform was transmitted from Chile to Japan, this does not mean that sea water (water particles) from the ocean off Chile was transported to Japan. Many people say this concept is difficult to understand.

To assist in understanding, let us consider a similar example. Image a soccer stadium full of fans. In the stands, which are filled with tens of thousands of people, a "wave" begins. Many readers have probably seen this in person or on television. The fans, with their hands raised, simply stand up slowly, and then sit down again. Everyone does this in order, at the proper timing. From a distance, it looks like a huge swelling, or wave, is circling the stands in a short time. Of course, the fans do not actually move around the stadium. Each fan only makes a few small movements at his or her own seat. The speed of this "wave" is decided by differences in the timing of the movements of each pair of persons sitting side by side. The wave travels quickly if the next person stands up immediately, and travels more slowly if the next person waits briefly before standing. However, the human sense is nearly the same in everybody, with no large variations in their movement. As a result, it seems as though the wave is transmitted at a roughly uniform speed. A tsunami corresponds to this "wave," and the movement of the water particles corresponds to the movement of the fans.

The velocity of water particles resulting from a tsunami, in other words, the speed with which the water particles move, is at maximum around 5 cm/s, assuming a tsunami with an amplitude of 1 m in the center of the Pacific Ocean. In a tsunami with a period of 20 minutes, the water moves in a reciprocating motion, advancing about 10 m and then returning to its original position in one wave period. ("Amplitude" is height from the center to the top or bottom of a wave; "period" is the time it takes for one complete wave to pass a given point.) The average movement velocity of the water is less than 2 cm/s, or only 72 m/h. Because the height of tsunamis propagating in the Pacific Ocean is considered to be several 10 cm, the speed of the water moved by a tsunami is at most several cm/s. This means that ships navigating the ocean scarcely notice a tsunami, even if they encounter one, and navigation is not hindered by tsunamis. Similarly, if a person were floating in the Pacific Ocean, the tsunami would not carry that person at the speed of jet plane (speed at which the tsunami waveform is transmitted), but rather, at the speed of the water particles, that is, several cm/s.

The amplitude of a tsunami is several meters at the tsunami source area, but this decreases as the tsunami propagates. The reason for this is as follows: Because a tsunami propagates from the source area in all directions, the propagation area increases as the propagation distance becomes longer, resulting in weakening ("attenuation") of the tsunami. This phenomenon is called "distance attenuation." Assuming propagation of a tsunami radiating from a source in water of uniform depth, attenuation is in inverse proportion to the square root of distance. In actuality, however, because the tsunami transmission speed varies in proportion to the square root of the water depth, as shown in Eq. (1), the propagation velocity of a tsunami is influenced by the topography of the sea bottom. Difference in the shape of the sea bottom (differences in depth) cause bending, or "refraction" of different parts of the wave. This means that tsunamis do not necessarily propagate in this radiating form, and attenuation is not always in inverse proportion to the square root of distance. Other types of attenuation are also caused during propagation of the tsunami by the viscosity of the water and friction with the sea bottom. Because attenuation due to viscosity and friction is larger in the short period fluctuation component, attenuation occurs from the short period component. This means that the short period component is gradually attenuated and the long period component tends to become predominant as the propagation distance increases. For example, tsunamis occurring near Japan have a period of around 10 to 40 minutes, but when the 1960 Chilean Earthquake Tsunami reached Japan, its period had increased to around 1 hour. As a result, even if a tsunami of the same height attacks the same bay, the damage will sometimes differ greatly in a tsunami originating in waters near Japan and a tsunami originating in a remote area such as the Chilean Earthquake Tsunami. This difference is caused due to the fact that the amplification of the tsunami depends on the natural period of fluid motion of the sea water in a particular bay. This phenomenon is called "resonance." When a bay is attacked by a tsunami that has a period close to its natural period, the tsunami is amplified after it invaded the bay, resulting in a major inundation disaster.

It should also be noted that the center of the tsunami source area is not necessarily identical with the center or focus of the earthquake. In many cases, the two are different. Therefore, the general practice is to establish the fault plane from seismic observation records of the earthquake, and then designate the tsunami source area by calculating the displacement of the sea bottom ground from the movement of the fault plane. On the other hand, when a tsunami is observed with tide gauges, it is also possible to locate the starting point of the tsunami by "inverse-propagation" of the tsunami, that is, by tracing back from the times when the tsunami arrived at various observation points. This is possible because the wave velocity of a tsunami is a function only of the water depth, as shown in Eq. (1). The starting point of propagation of the tsunami is designated by this type of inverse-propagation calculation from each of the points where the tide (tsunami) level was measured, and the region enveloped by various starting points thus calculated is considered to be the tsunami source area.

(2) *Tsunamis in coastal areas (water depth 30 m to 2 m)*

Assuming the eye height of a person standing on the shore is 1.8 m above sea level, that person can theoretically see to a distance of about 4.8 km offshore. Because the earth is round, it is impossible to see further than this. Actual visibility is probably no more than 3 km. A Japanese children's song says "The ocean is wide, the ocean is big," but the ocean that people can see from the shore is definitely not large. This section will describe the deformation of tsunamis in the narrow range of ocean that people can see. In terms of water depth, this is shallow waters with a depth up to 30 m, assuming the seafloor slope is 1/100.

Even in shallow water, the velocity at which the tsunami waveform propagates can be calculated by Eq. (1), just as in deeper water. Although the propagation velocity of tsunamis is comparable to the speed of a jet plane in the open ocean, this speed decreases steadily as the tsunami approaches shore, from 62 km/h at a water depth of 30 m to 50, 35 and 25 km/h at depths of 20, 10 and 5 m, respectively. This is the speed of an automobile, and closer to the shore, the speed of a bicycle. On the other hand, as the water depth becomes shallower, the height of the tsunami increases rapidly, in inverse proportion to the 1/4 power of the water depth. The velocity of the water particles also increases. At the same time, the waveform also changes greatly, with the wave front forming a steep slope. A wave with a steep slope will deform in an attempt to stabilize, but subsequent changes in the waveform will differ depending on the slope of the sea bottom.

When the seafloor slope is extremely gentle with 1/200 or less, small waves are generated on the surface of the water in an effort to maintain the steep-sloped waveform on the front side of the tsunami wave. These small waves have a period similar to that of ordinary wind waves, i.e., 7-8 s. These waves ride on the main body of the tsunami and are gradually amplified by energy supplied from there. The phenomenon by which these short period waves appear is called "soliton fission." As these short period waves develop, new short period waves are generated behind them, and the number of short period waves increases. When short period waves reach a certain size relative to the water depth, they break and quickly attenuate. Then, new waves develop next, and when they achieve the conditions for breaking, they too break and quickly decrease in size. This successive breaking during propagation continues until finally only the main body of the tsunami remains. Because the energy of the tsunami is used in the development of short period waves, and this energy is lost when the short period waves break, the breaking of the short period waves reduces the size of the main body of the tsunami. If the slope becomes steep while these short period waves exist, the short period waves will break suddenly on the steep slope. Because the main body of the tsunami will still remain, even in this condition, the tsunami will be reflected by the coast toward the open sea, regardless of whether the seafloor slope is gentle or steep. If people are engulfed in breaking short period waves, they will be violently tossed about by the breaking waves, become unable to control their situation, and may drown.

When the seafloor slope is somewhat steeper, in the range of 1/50 to 1/100, the size of a tsunami will increase due to "shallow water deformation" as the wave advances toward shore and the water becomes shallower. The front side of the tsunami wave becomes steeper, causing soliton fission, but the waves will break immediately near the shoreline. In this case, the soliton fission waves break together with the main body of the tsunami. People who are engulfed under these conditions are in extreme danger. When the seafloor slope is steeper than this range, the tsunami will be completely reflected without breaking, and will display the behavior of a standing wave. Under these conditions, there is a high possibility of survival if a person can grasp some floating object. Moreover, it is also possible to hold onto an object under this condition.

The following will present experimental results showing the form of actual tsunamis.

Figure 2.2(a) shows a wave channel which is used to generate artificial tsunamis. In this 50-m long channel, the front slope rises at 1/10, after which the channel is connected to a reef of uniform water depth. The length of the reef is 20 m. Near the shore line, the channel simulates a coast with a uniform slope of 1/20. In the experiment, a

10-cm model seawall is placed on the coast at a position 2 cm above the still water level (SWL). Supposing a tsunami attack on a reef-type coast of this kind, an experiment was performed using various water depths at the reef. If the scale of this model experiment is considered to be 1/100,

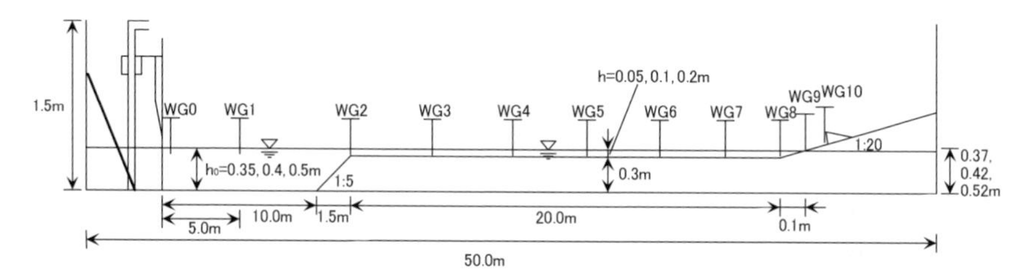

(a) Simulated topography in tsunami channel

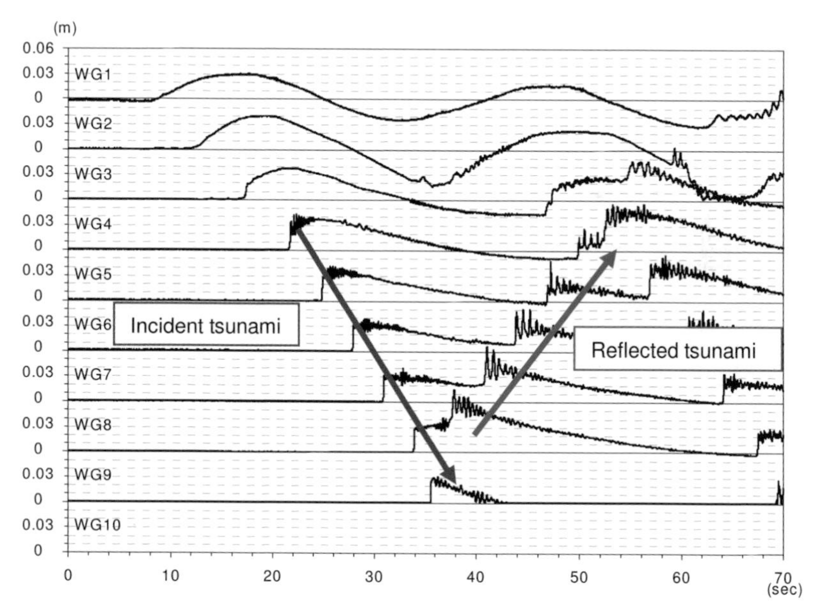

(b) Tsunami deformation in a shallow sea of 5 m (at actual scale: η_0 = 3 m, T = 5 min, H = 5 m)

Fig. 2.2 Deformation of the tsunami waveform depending on the relationship between tsunami height and water depth.

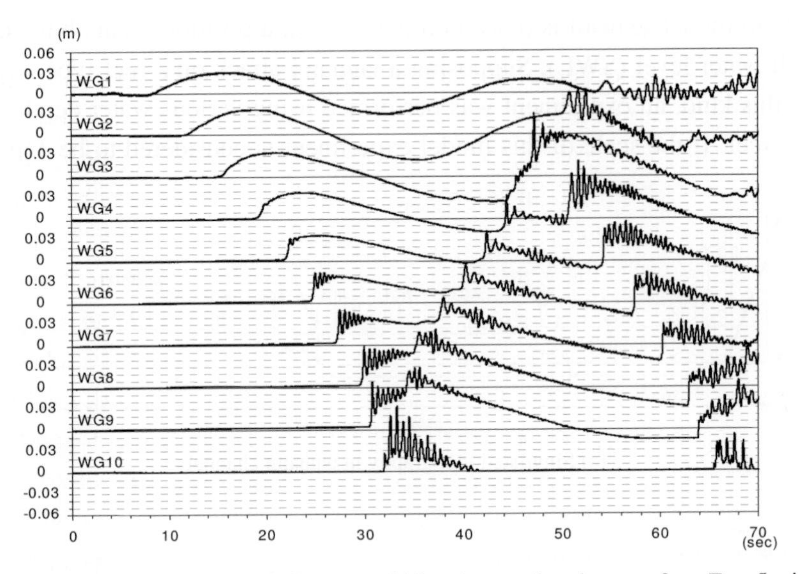

(c) Tsunami deformation in a shallow sea of 10 m (at actual scale: η_0 = 3 m, T = 5 min, h = 10 m)

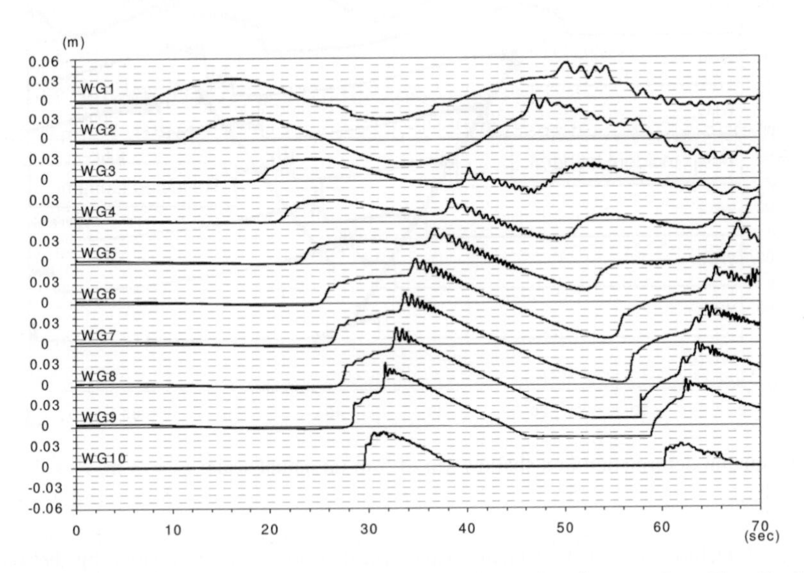

(d) Tsunami deformation in a shallow sea of 20 m (at actual scale: η_0 = 3 m, T = 5 min, h = 20 m)

Fig. 2.2 (*Continued*)

the length of the actual reef would be 2 km, and the various water depths at the reef would be 5, 10, and 20 m. The height and period of the attacking tsunami were set at 3 m and 5 min, respectively. The waveform of the tsunami was measured with 9 to 10 units of wave gauges (WG) positioned on the reef. In the following, the results are described using prototype scale values.

In Fig. 2.2(b), the water depth on the reef is 5 m. When the tsunami runs up on the reef, the height of the tsunami increases from 3 to 4 m due to shoaling deformation as shown by wave gauge WG2, which was installed at the shoulder of the reef. As it propagates on the reef, the front waveform becomes steeper, and new short period waves are formed by soliton fission. They are generated by the effects of nonlinearity and dispersion, which are strengthened when the waveform of the tsunami becomes steeper. Their period is 7-8 seconds and basically the same as that of ordinary waves. The leading wave is the highest, and those that follow become progressively smaller. However, due to the shallow water depth, the wave height rapidly becomes smaller due to successive breaking of the waves, beginning with the first wave. Finally, in front of the seawall, all the short period waves almost vanish by breaking, and the main body of the tsunami is also reduced to 3 m (WG 6-9). Even if this tsunami runs up on land, it will not cause extremely large damage. The condition of the tsunami as it returns to sea after being reflected by the seawall can be clearly seen in the figure. When the reflected tsunami wave collides with the crest of the incoming ("incident") tsunami wave at WG 6-8, large soliton fission waves are remarkable on the reflected Tsunami. However, when the reflected wave meets the trough of the incident wave at WG 4-5, the size of the soliton fission waves is reduced as a result of breaking due to the shallow water depth.

Figure 2.2(c) shows the case when the water depth on the reef is 10 m. Unlike the 5-m case, after the tsunami runs up on the reef, the front waveform becomes steeper as the wave advances. Fission begins when the tsunami has advanced several 100 meters across the reef (WG5), and the number of short period fission waves increases as these waves continue to develop (WG 6-9). Because the largest of these fission waves is the 1st wave, and the 2nd, 3rd, and later waves become progressively smaller, the 1st wave breaks first. After the 1st wave breaks, the 2nd wave breaks. Due to the deep water depth in this 10-m case, these short period waves do not break unless their wave height exceeds 7 m. Therefore, the waves break upon reaching the water on the 1/20 slope near the shore line. Because this tsunami overtops the seawall, the tsunami is only partially reflected back to sea. The condition of the reflected tsunami wave returning to the sea can be seen in the figure. Breaking tsunami in the state of soliton fission exerts very large impulsive force on a seawall, and a person caught by the breaking tsunami is brought to death with very high possibility.

Figure 2.2(d) shows the case when the depth of the water on the reef is increased further, to 20 m. When the water is this deep, the increase in tsunami height by shallow water deformation is slight, even when the tsunami runs up on a reef. Although the tsunami becomes forward-leaning and its front slope becomes steeper on the reef, no clear soliton fission occurs. The absence of fission is explained as follows. When the water depth is deep relative to the height of a tsunami, a long reef is necessary for fission. However, in this experiment, the reef was short (2 km). If the reef were sufficiently long, it is thought that soliton fission would occur. With this type of tsunami, wave-breaking occurs all at once at a coast having a slope of 1/20. Thus, even if a reef exists, soliton fission occurs only when the ratio of the tsunami height versus water depth is large, and does not occur when the ratio is small and the reef is short. Thus, the difference in the tsunami deformation generates a large difference in the tsunami force acting on coastal structures and as such, has a large effect on damage to buildings and human life.

The sea area where the deformation of a tsunami is caused as explained here is within the range people standing on the shore can see. In other words, it is an area where people on shore can see a tsunami attack with their own eyes. However, can people recognize a tsunami attack simply by looking at the sea, and do they begin evacuation without having any knowledge of tsunamis and without receiving any information such as a tsunami warning? Unfortunately, this is extremely difficult. Although a tsunami undergoes rapid deformation in the shallow water depth, this begins at a distance of several kilometers from the shore, which is too great for people standing on the shore to recognize the attack of the tsunami. If people happen to be looking out to sea, they may sense something is strange about the ocean surface. However, they will not become aware of the most important feature of a tsunami; that is, the fact that the swelling of the sea extends over an extremely wide area behind it. Moreover, because this change in the sea seems far away, the feeling of curiosity ("what's that?") may be stronger than the feeling of fear.

Now, let us assume that a tsunami appears 3 km offshore, which is approximately the limit that a person can see from shore. Assuming the slope of the sea bottom is 1/100, the water depth at a distance of 3 km will be 30 m. The time necessary for the tsunami to cross this sea bottom slope to the 2-m deep water area just in front of a person standing on the beach can be calculated using Eq. (1). The tsunami will arrive in only 4 minutes and 20 seconds. A person can easily waste time by hesitating to flee. In other words, if you sense something is strange about the sea, you cannot hesitate for even an instant; you must flee immediately. During the Great Indian Ocean Tsunami, a little British girl noticed something strange about the ocean and immediately told those around her, who all fled together. In this case, these few minutes were the difference between life and death.

A survivor of the Great Indian Ocean Tsunami who was actually caught in the tsunami said that he did not feel frightened until the tsunami was almost on top of him. As a result, he fled too late. Figure 2.3 shows people watching the tsunami as it is going to attack them. They seem to think that a larger-than-normal wave is approaching, but have no idea what is actually happening. In other words, they have absolutely no sense that the swelling of the sea surface continues for long distance behind the wave front. As a result, they still have not begun to evacuate.

Fig. 2.3 People watching a tsunami without fleeing (Thailand, Great Indian Ocean Tsunami).

There are certain facts that you must know in order to escape from a tsunami: A tsunami attack on a coast may not be limited to only one wave, but may include several waves. When the displacement of the sea surface caused by a change in the sea bottom topography due to an earthquake begins to propagate as a tsunami, it generally consists of only one wave. The expression "generally" is used because, strictly speaking, this is not one wave, but is followed by a series of smaller waves. However, these small waves are extremely small in comparison with the first wave and can be ignored. This might suggest that a tsunami which is transmitted to a coast should consist of only one wave. In fact, however, an actual tsunami attack consists of multiple waves, and the 1st wave is not necessarily the largest. It is not uncommon that the 2nd or 3rd wave is even larger. For example, in the Great Indian Ocean Tsunami, many people fled the smaller 1st wave and survived, but died in the larger 2nd and 3rd waves.

The reasons why tsunamis attack in multiple waves are as follows: Firstly, the speed of a tsunami varies depending on the water depth. The speed of tsunamis is affected by the water depth, even in the deepest waters on the planet. The speed at which a tsunami propagates increases with water depth, as expressed by Eq. (1). This means that the 1st tsunami wave that arrives at an object coast is the wave that propagated selectively by the route with the deepest water depth, and the waves that propagate by other routes will arrive later. Regardless of which route the wave takes, the propagation velocity differs depending on the water depth. Therefore, a tsunami is affected by the sea bottom topography during propagation, and refracted (or bent) toward relatively shallower waters and the energy of the tsunami tends to concentrate in some parts and disperse in others. As a result, the 1st wave will not necessarily become a large wave. The second reason why tsunamis attack in multiple waves is that tsunamis may be largely reflected from even ordinary coastal lines. Reflection occurs because the wavelength of a tsunami is long, exceeding 10 km, and the wave height is small in comparison with this long wavelength. A reflected tsunami returns toward the open sea. Because tsunamis have the property of being bent toward shallower waters, there are cases in which a tsunami that has

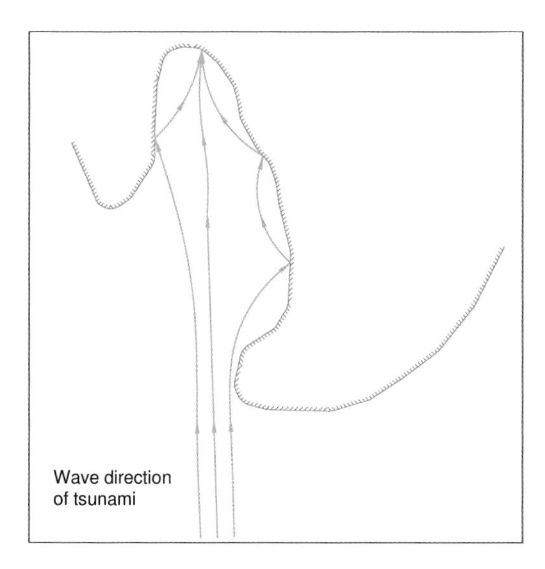

Fig. 2.4 Tsunami waves attacking by different routes.

been reflected to sea is bent toward shallower waters, and eventually returns and attacks a different part of the coastal line from the coast where it was reflected. If the conditions for repeated occurrence of this process exist, a tsunami with a particular period corresponding to the sea bottom topography may be trapped in the coastal sea area. Waves that are thus trapped in the coastal sea area are called “edge waves” because they occur at the edge of the ocean.

Figure 2.4 illustrates the phenomena of refraction and reflection. As shown in this figure, the waves which occur at the object coast include the wave that approaches the coast directly, waves that approach different sections of the coast as a result of refraction, waves that approach the coast after being reflected, and in addition to these, edge waves. Because these are superposed with some deviation in their phases, multiple waves of the tsunami will attack the coast.

(3) *Tsunamis on land areas (water depth 2 m to land areas)*

On land, tsunamis are affected directly by the topography of the land and change greatly as a result. The effect of land topography will be explained using the three typical cross sections shown in Fig. 2.5.

First, Fig. 2.5(a) shows a cross section with a uniform slope. The run-up height of a tsunami on this cross section was obtained from a hydraulic model experiment, as shown in Fig. 2.6. The x-axis in this figure is the ratio of the horizontal length l of the underwater part in the uniform slope to the wavelength L of the tsunami, and shows the effect of the sea bottom slope and period of the tsunami. The wavelength of the tsunami is a value for the water depth at the point where the uniform slope begins. The y-axis shows the ratio of the tsunami run-up height R and the tsunami wave height H. The tsunami run-up height is the vertical height of the highest point reached by the tsunami measured from the still water level. The tsunami wave height is given as the wave height at the leading edge of the uniform slope. Assuming a tsunami with a period of 10 min attacks a coast having a uniform slope of 1/50 from the point where the water depth is 30 m, the horizontal length of the uniform slope is l = 1,500 m and the wavelength of the tsunami is L = 10,300 m; therefore, $l/L = 0.15$. From Fig. 2.6, the run-up height ratio is $R/L = 3.0$. This means, for example, that a tsunami with a height of 5 m will run up to a height of 15 m above the still water level.

The maximum horizontal velocity of tsunamis on land can be roughly estimated from the following equation.

$$u = A\sqrt{g(R - h_G)} \tag{2}$$

where, u is the maximum horizontal velocity at a point with the height above sea level h_G. The value of A can be considered to be on the order of 1.0. Although the run-up height R can be obtained from Fig. 2.6, as a conservative value, this can be regarded as three times the wave height of the tsunami. Thus, according to this equation, a 5-m tsunami will have a maximum flow velocity of approximately 7 m/s at a point 10 m above sea level.

Next, in the cross-section in Fig. 2.5(b), the uniform slope becomes horizontal at the point with the height above sea level h_G. In this case, if the run-up height exceeds the height h_G, the tsunami will run up on land until it reaches deep inland. The flow velocity on level ground can be calculated by the above Eq. (2). However, on horizontal ground, the flow velocity will be gradually reduced by friction between the sea water and the ground as the tsunami advances inland.

Figure 2.5(c) is the case in which land with a uniform slope becomes an reverse (downward) slope in the landward direction at the point with the height above sea level h_G. If the run-up height exceeds height h_G, the velocity of the tsunami will be accelerated by the downward slope in the landward direction. At coasts with this shape, extremely heavy damage can easily be imagined. In the 1983 Nihonkai-Chubu Earthquake Tsunami, the tsunami which ran up on the coast crossed sand dunes 12-13 m in height and then ran down the back side of the dunes, killing women working in rice paddy. In the Great Indian Ocean Tsunami, the location where the train was overturned at Kahawa, Sri Lanka also had this kind of topography (see 1.2(5)).

As described above, the behavior and flow velocity of tsunamis on land is strongly affected by the land topography. The tsunami may become either stronger or weaker, depending on the topography. In particular, when the land is flat, the flow velocity tends to decelerate very slowly, and the tsunami can reach areas deep inland. When the land has a downward slope toward the landward side, the tsunami flow

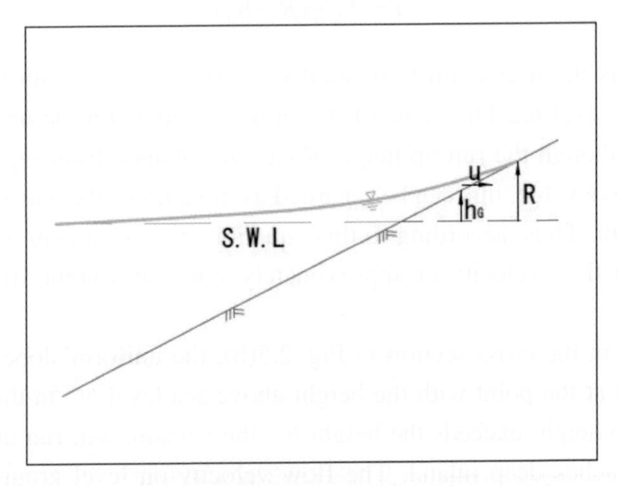

(a) Topography with uniform slope

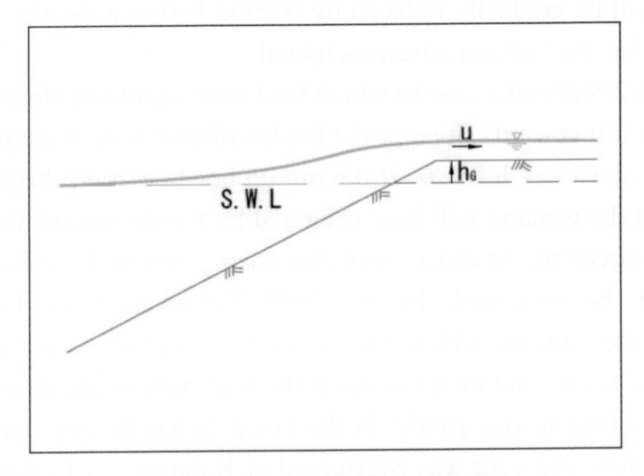

(b) Topography with horizontal slope

Fig. 2.5 Typical coastal topographies.

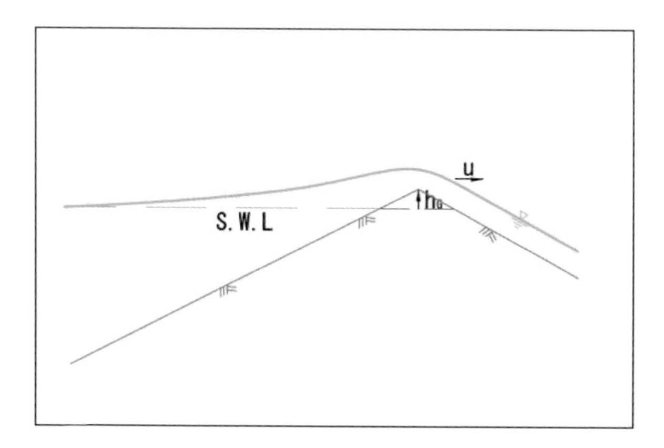

(c) Topography with reverse slope

Fig. 2.5 (*Continued*)

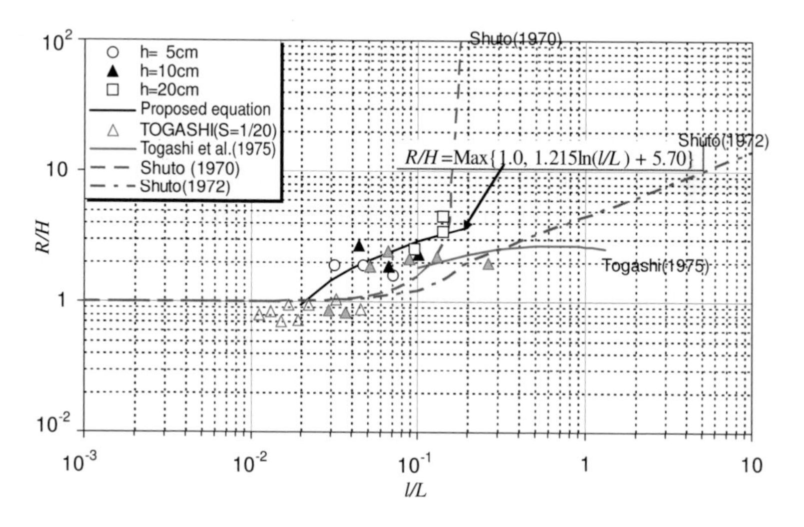

Fig. 2.6 Tsunami run-up height.

is accelerated and the flow velocity becomes extremely high. Therefore, when the land is flat or slopes downward toward the landward side, people may be engulfed in the tsunami if they do not evacuate quickly to high ground or a suitable high building.

References

Aida, I. (1981): Numerical Experiments of Historical Tsunamis Generated Off the Coast of the Tokaido District, Bulletin of the Earthquake Research Institute, University of Tokyo, 56, pp. 367-390. (in Japanese)

Matsutomi, H. and Iizuka, H. (1998): Tsunami Current Velocity on Land and Its Simple Estimation Method, Annual Journal of Coastal Engineering, Japan Society of Civil Engineers, Vol. 45, pp. 361-365. (in Japanese)

Yasuda, T., Takayama, T., and Yamamoto, H. (2006): Characteristics of deformations and forces on dispersive soliton tsunami, Annual Journal of Coastal Engineering, Japan Society of Civil Engineers, Vol. 45, pp. 256-260. (in Japanese)

Yasuda, T., Takayama, T., and Yamamoto, H. (2006): Effect of coastal cross section on characteristics of tsunami deformations and forces, Annual Journal of Civil Engineering in the Ocean, Japan Society of Civil Engineers, Vol. 22, pp. 529-534. (in Japanese)

2.2 Tsunami Currents and the Danger to Humans

The period of a tsunami wave is extremely long. Therefore, the flow caused by a tsunami is more like a river than an ordinary wind-generated ocean wave. However, unlike a river, the depth of oceans changes from several thousand meters to 0 m at shore. As a result, the flow of water in a tsunami also changes greatly, depending on the water depth. For example, when the ocean depth is 1,000 m and the change in water level caused by a tsunami is ±0.5 m (1 m in wave height), the speed of the flow of sea water, in other words, the flow velocity, is at most only 5 cm/s. However, as the tsunami approaches a coastal line, where the water depth is shallow, the change in the water level becomes extremely large, and the flow velocity of the tsunami increases corresponding to the change. When the water depth is 10 m and the vertical change in the water level is ±1.5 m (3 m in wave height), about three times larger than that at 1,000 m deep, and the maximum flow velocity is around 150 cm/s. Closer to the coastal line, where the water is very shallow, the size of the tsunami increases suddenly, the flow becomes faster, and the leading edge of the tsunami begins to break. The speed of the leading edge of the tsunami is fast, even when the water depth is shallow. When the wave breaks, the crest strikes the water surface, causing violent turbulence. Finally, the tsunami runs up on land while continuing to break in this violent manner.

One reason for the large loss of life in tsunamis is the flow near the shore accompanied by violent turbulence at the leading edge, and the unique flow that follows the leading edge. A tsunami wave and an ordinary wave break in a relatively similar way. People who live near the coast or those who have experience of swimming in the ocean have seen ordinary waves breaking and can probably understand how a tsunami wave breaks at its leading edge. However, the decisive difference between a tsunami and an ordinary wave is the extremely long period of a tsunami wave. Even the longest ordinary waves will retreat after about 10 seconds. However, in a tsunami, the flow continues to come for a long period of time even after the leading edge of the wave passes. As a result, sea water may reach inland areas as far as several km from the coast. A few to several ten minutes may pass before the tsunami waters finally begin to return to the ocean. The force of this return flow is far stronger than one might expect, and can carry large objects on land into the sea. This kind of tsunami wave engulfs many people and can cause deaths.

(1) *Dangerous locations and conditions*

Table 2.1 summarizes the direct danger to the human population in a tsunami. The greatest danger to people is (A) being caught in the breaking part of the wave, followed by (B) falling down as a result of the fast flow. Persons who fall in the wave can be injured, and if they lose consciousness, they may drown. Next, (C) if the water is higher than a certain level, people may float rather than fall. As a result, they may lose their balance and, if their feet cannot touch the ground, they may drown.

In the following, the term "water depth" means the water depth before the tsunami attack when it is referring to a sea area. On land, the water depth is normally 0; therefore, "water depth" means the inundation depth due to a tsunami.

Table 2.1 Calculation of human danger due to tsunamis.

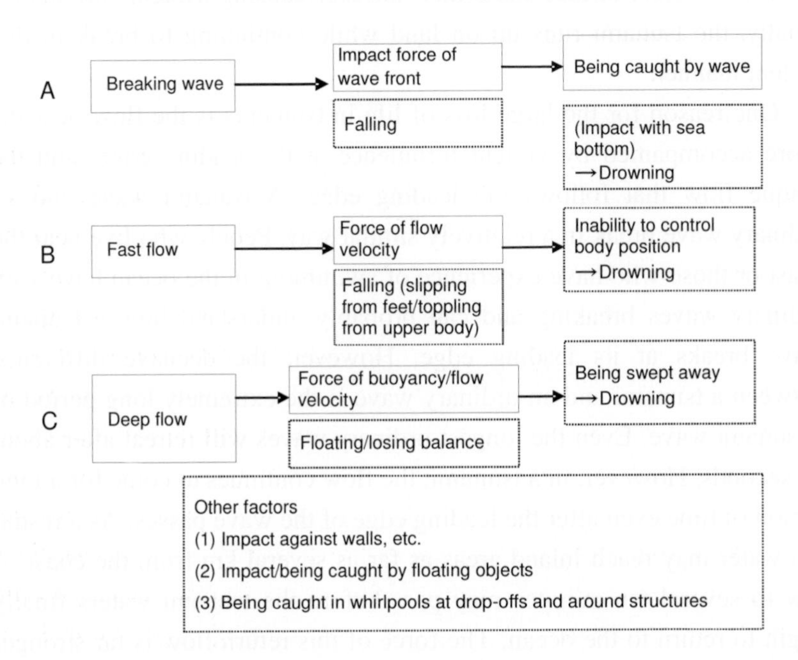

(1) *Dangers at and around the shore*

The most dangerous area in a tsunami is the area around the shore (beaches and other shores where the wave strikes), as shown in Fig. 1.1 in 1.1. Near the coastal line, both the speed and the turbulence of the tsunami flow increase, and even good swimmers are unable to help themselves. These conditions cause many deaths. Among the dangers near the shore, when the tsunami wave breaks and the wall of water hits people, they may be dragged to the sea bottom by the vortex of the breaking wave and drown, or they may be thrown against the sea bottom and injured (Table 2.1, A). After the wave breaks, a rapid flow continues. Even in shallow water (less than waist deep), this flow can cause people to fall, be caught in the flow, and drown (Table 2.1, B). There is also a danger of injury due to being thrown against the sea bottom or other hard surfaces, and another danger of being caught in the debris from houses and other objects destroyed by the tsunami. As the water depth increases, the danger of falling also increases, and people may fall under the water, lose their balance, and drown.

(2) *Dangers at some distance from the shore*

Figure 1.2 in 1.1 showed a bus terminal a short distance from the coast. Here, the water depth was 1-2 m and the flow velocity was 1-2 m/s or more. People were swept away by this flow. Due to the distance from the shore, the effect of the breaking wave was slight and the flow velocity had decreased. However, even with a flow velocity like that in the figure, there is considerable danger of falling. In particular, when the water level is chest high or higher, the effect of buoyancy (floating) must be considered. In deep water like this, it is possible to lose your balance and be swept away, even when the flow velocity is not especially fast. If the flow velocity is not very fast, swimmers can stay afloat without falling, but faster currents can engulf even good swimmers. These conditions correspond to B and C in Table 2.1. As can be seen in the figure, the tsunami flow contains various floating objects such as debris from houses which were destroyed in the waterfront area. There is a high danger of being caught in these objects.

(3) *Dangers where flows become faster*

In general, the shape of the land ("topography") is complicated. Therefore, the tsunami flow velocity and water depth are not uniform. Places with deep water and rapid flows are extremely dangerous. For example, the back side of sand dunes at the seacoast has a so-called "reverse slope" (downward slope). The flow accelerates down this slope and becomes extremely dangerous. When a lagoon exists in the hinterland, a reverse slope is formed. The tsunami will naturally flow into these low spots, causing whirlpools and other severe turbulence. In the past, many deaths were caused in these circumstances. On the other hand, when the land forms a long, narrow cape, like the Aonae District in the Hokkaido Nansei-oki Earthquake, the tsunami may cross the cape from one side to the other. In this case, many deaths were caused by the powerful flow on the opposite coastal line away from the tsunami attack.

Tsunami flows also concentrate in watercourses where water normally flows, such as rivers, smaller streams, drainage channels, and the like. In particular, the flow concentrates in these watercourses during the return flow from the tsunami, generating extremely fast flows. Thus, the danger of drowning is high.

(4) *Return flow*

After inundating the hinterland, a tsunami will eventually return to the sea. This is called the "return flow." This return flow does not endanger people with a breaking wave. However, due to the combined conditions of water depth and flow velocity, the danger of falling and being caught in the flow is the same as during run-up. In general, the velocity of the return flow is slower than that of the up-flow, and the danger of falling is comparatively smaller. However, the return flow generally flows down a slope as it returns to the sea. The flow can become extremely fast as it approaches the coast, and can drag people into the sea. Therefore, care is necessary. At the coastal line, the land may be steep and there may be sudden drops, for example, at embankments. There is a danger of falling from these places and being caught in the tsunami flow. This can result in drowning, injury by being thrown against the sea bottom, etc.

As a distinctive problem during the return flow, people may catch some floating object and be carried by it. It may be necessary to catch some floating object in order to survive. In fact, in many cases people have survived while being carried by the wave after catching a floating object. Sometimes people are carried away offshore by the return flow. In these cases, many people have been discovered immediately after the tsunami and saved. However, some unlucky people are carried great distances by ocean currents and spend a long time in the water. Exhaustion to death are quite possible, especially in cold seasons.

(2) *Impact of breaking tsunami wave on humans*

Direct impact by the wall of water at the wave front of a breaking tsunami (Table 2.1, A) is extremely dangerous to humans. This section will describe the mechanism, the danger to humans, and methods of escaping.

Figure 2.7 shows a wave breaking beautifully a short distance offshore. Many people who enjoy swimming in the ocean have had the experience of being hit by the front of a wave just before the wave breaks, as illustrated in Fig. 2.8. In large waves, people may be knocked downed, rolled along in the water, or thrown against the sea bottom. The most dangerous situation is when a person is hit directly from behind by a wave that is just starting to break. Such persons may be knocked out (lose consciousness) or suffer serious injury if the sea bottom is rocky. When a tsunami wave breaks and strikes and catches people, the mechanism is basically the same. However, because ordinary waves retreat in only a few seconds, people who are caught in the wave can usually stand up in the water. In contrast, tsunamis do not retreat immediately, and the danger of drowning is high.

When a tsunami strikes a person, impact force acts on that person. The impact time is short, but the force is very large. The force varies greatly depending on the posture of a person at the moment of impact. The effect is similar to diving into a swimming pool. For example, poor divers smash into the water, their entire chest and belly striking the water with enormous force. Good divers enter the water smoothly with hardly a splash. In other words, when the body is parallel to the water, the diver hits the water with great force, but as the angle of the body increases, the impact force becomes smaller. In a research on the characteristics of impact force, the force of waves acting on a round post standing on the shore was investigated. However, the impact on humans is more complicated than that on fixed posts, because people can move their bodies and can be moved by impact force. For this reason, the effect of the impact force of waves on human beings is still not adequately understood. A rough safety limit obtained from past experiences at several coastal lines is when a wave is beginning to break or is breaking and the height of the crest is just over a person's head. Beyond this limit, danger increases with increasing wave height.

Fig. 2.7 Swimmers and breaking waves.

Fig. 2.8 Impact of the breaking wave front on person and falling.

Whether a tsunami breaks or not, and where the wave breaks, depend on the height of the tsunami and the slope of the sea bottom. Conditions that result in breaking frequently exist near the shore, and accordingly, the danger is high. The location where tsunami waves break, and the height of the wave when it breaks, can be obtained by referring to the results of numerical calculations on tsunami run-up. In many cases, breaking tsunami waves destroy buildings and other structures, and are also a problem when considering refuge areas.

In any case, in order to avoid a breaking tsunami, it is essential to evacuate quickly from beaches and other waterfront areas. If you are on land, you can evacuate if you realize the danger quickly. However, if it is too late to evacuate, you should take shelter behind a sturdy building nearest to you. If you are in the water and it is too late to escape, at least try to avoid being hit by the wall of water or caught in the breaking wave. For example, surfers take shelter in the bottom of the wave to avoid the breaking wave. This could also be one of the ways available for you to avoid the breaking wave and survive.

In Fig. 2.7 showing the Krabi coast of Thailand, the tsunami is causing "soliton fission." Soliton fission is a phenomenon that can be seen at a coastal line with a gently sloping seabed, where wave motion like that in ordinary waves occurs at the leading edge of a tsunami. These soliton fission waves move forward riding on the main body of the tsunami and then break like ordinary waves. As a result, the breaking of the individual soliton fission waves may not be particularly violent. It is presumed that this is the reason why fortunately, many of the people shown in the figure survived the tsunami, even though they were hit by these breaking waves while they were in the water.

(3) *Falling due to the force of flow velocity*

In Table 2.1, Case B shows the danger of a person losing balance and falling due to the force of the tsunami flow. If a person loses his or her balance and is unable to stand again in the flow, the danger of drowning is high. Victims may also be caught in debris being carried by the same flow. In this case, they may be injured or unable to move because of the debris, again resulting in drowning. First, we will consider the conditions that cause people to lose their balance and fall due to the forces in a tsunami flow. We will also explain methods of avoiding this kind of danger.

The fluid force (or drag force, F) acting on a person due to a tsunami flow is proportionate to the second power of the velocity of the flow, U. This force causes people to lose their balance and fall. Figure 2.9 is a schematic illustration of the mechanism by which people fall in a tsunami flow. Basically, the resistance to the flow force F is equivalent to the weight of the person's body (strictly speaking, the weight W_0 after subtracting buoyancy). As shown in the figure, there are two ways that a person can lose his or her balance and fall, (1) slipping (from the feet) and (2) toppling (from the upper body).

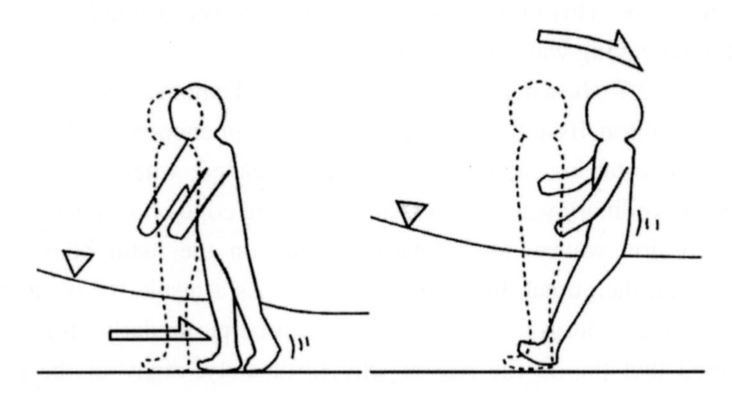

Fig. 2.9 Schematic illustration of mechanism of falling in a tsunami flow (source: Takahashi et al., 1992).

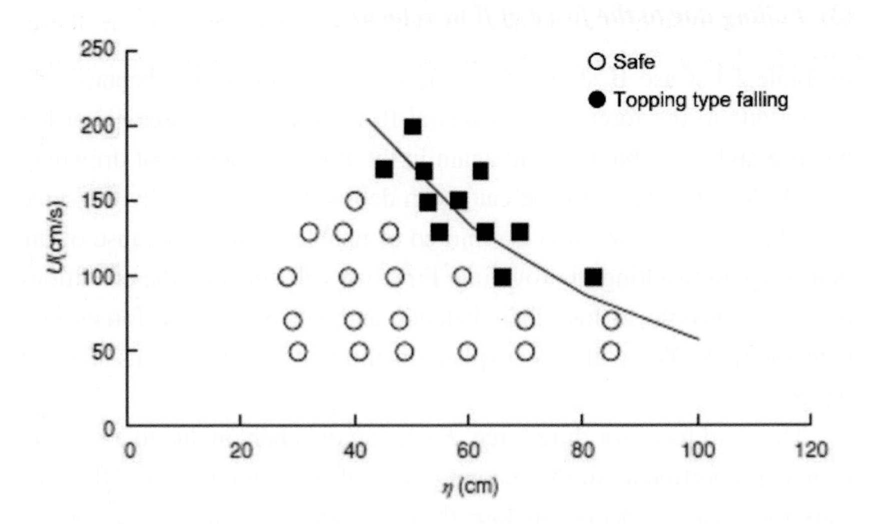

Fig. 2.10 Results of an experiment on falling in flow (source: Takahashi et al., 1992).

(1) In slipping, the person loses balance and falls from their feet when the fluid force (drag force, F) exceeds the frictional resistant force ($\mu_s W_o$) of the soles of their feet due to body weight. (Here, μ_s is the coefficient of friction.)
(2) In toppling, the person falls over from the upper body when the moment (Fh_G) around the point S (bottom back of the heel) due to fluid force (drag force, F) exceeds the resistance moment ($W_o l_G$) due to body weight. (Here, h_G is the height from the ground surface to the center of gravity of the body, and l_G is the distance from the center of gravity of the body to the bottom back of the heel.)

Figure 2.10 shows the results of an investigation of human stability against flows in a water channel experiment. The subject in this experiment was a thin, slightly tall (183 cm) person. The subject lost his balance and fell in slipping type falling (in which his feet slipped out from under him) at a water depth of 60 cm and flow velocity of approximately 1.2 m/s, and at a depth of 40 cm and velocity of 1.8 m/s. Normally, it is possible to keep one's balance by changing your stance and moving your feet and upper body. However, in order to obtain conservative results, the subject in this experiment did not use these methods. A flow velocity of 1-2 m/s is about the same speed as walking. This is not an especially fast flow velocity in a tsunami. In fact, this flow velocity can occur anywhere during run-up of the tsunami. This means that there is a very high danger of losing your balance and falling in a tsunami.

Figure 2.11 shows the relationship between the water depth η and the critical velocity for losing one's balance, U_{cr}. These are calculated results for a 160-cm tall person. The solid lines show (1) slipping type falling. U_{cr} differs depending on the coefficient of friction μ_s of the soles of the feet. The broken line shows (2) toppling type falling. When the water level is a certain height (approximately higher than the thighs), the falling mode becomes toppling type falling. Because the subject was a shorter person than in the experiment in Fig. 2.10, and the values reported in Fig. 2.10 were intentionally conservative, the critical flow

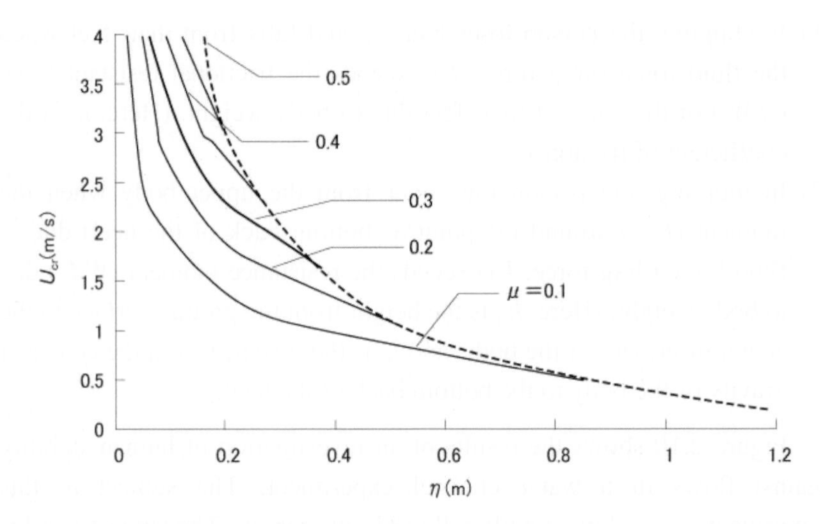

Fig. 2.11 Critical flow velocity U_{cr} for falling due to a flow, for a person with height of 160 cm (source: Takahashi et al., 1992).

velocity is somewhat smaller in Fig. 2.11 than in Fig. 2.10. In any case, there is a danger of losing your balance and falling even with relatively small flow velocities. For example, a flow velocity of 1.3 m/s can cause falling when the water is only knee-deep, and a velocity of 0.6 m/s can cause falling when the water is at mid-body level.

As mentioned previously, if you lose your balance and fall in a tsunami flow, you are in an extremely dangerous situation, because it is very difficult to stand up and regain your balance. People who fall in a tsunami are often carried away in the same position as when they fell. For example, a person who was sitting in a chair was carried away, still in the chair, to deeper water more than 30 m away. Research on what happens after falling is still inadequate. However, it is clear that one's position when falling often means life or death. At minimum, it is necessary to avoid being face down after falling.

As mentioned earlier, the water level and flow velocity at various locations can be obtained by numerical calculations of the tsunami run-up. In particular, it is essential to know which locations will have locally high flow velocities.

(4) *Flows with deep water depth*

When the water is over your head, or above a certain level (higher than your chest or neck), as in Case C in Table 2.1, you will lose your balance due to buoyancy. As a result, you will float rather than fall. If the flow velocity is not especially high and you can swim, you can float and be carried by the flow. You may actually be safer than in shallow water.

However, most people who can swim cannot fight a strong, fast current. Even trained swimmers have difficulty in swimming for long periods when the current is more than 0.5 m/s, and it is quite hard to fight a current of only 0.3 m/s. Swimming is almost impossible in very fast currents and when a whirlpool forms. It is necessary to save your strength, limiting your actions to finding some floating object to support you, avoiding dangerous locations, and searching for a place where one can secure safety.

(5) *Distance victims are carried by return flow*

In the tsunami return flow, as in the up-flow, human stability in the flow depends on the flow velocity and the water depth. During the return flow, it is especially important to avoid being carried into dangerous locations. For this reason, it is important to think about the route that the flow will carry you. Detailed information can now be obtained by numerical calculations. Here, however, we will use simple calculations to show the distance that you might be carried. This distance is greater than you might expect.

The distance of movement R to the landward side or seaward side due to a tsunami is $R = HL/(4\pi h)$, where h is the water depth, H is the height of the tsunami, and L is the wavelength. Because this assumes a certain water depth, strictly speaking, it cannot be used on land. However, it can be applied to land in an approximate manner. For example, assuming a wave period, T of 30 minutes and a water depth, h of 4 m, the wavelength, L of the tsunami is 11.27 km. If the wave height is 1 m, R = 224 m. As the period and height of the tsunami increase, R also becomes larger. Considering this, a large tsunami which inundates the inland area can carry you over a distance of several hundred meters to several kilometers in the landward direction. Likewise, when the tsunami returns in the seaward direction, there is also a high possibility of being carried into the ocean, and then carried by tidal currents or other ocean currents.

(6) *Protecting yourself from tsunami flows*

In order to protect yourself from tsunami flows, first, you must recognize that "tsunami flows are extremely dangerous, and it is extremely important to avoid them." In particular, you must understand that evacuation from the dangerous coastal area can mean the difference between life and death. After a warning is issued, you must not go to a port or the coast to see the tsunami. Even in shallow water, your legs can be caught by a tsunami with ordinary walking speed. In faster flows, you may be helpless, even if you can swim. In particular, there is a danger of being caught in the tsunami flow by a whirlpool or the like. If you have the bad fortune to be caught in the water, do not give up. You must use all your efforts to survive, for example, by moving to the highest area possible, behind a building, or to some other place where the flow is relatively weak.

If you are a little late in evacuating, sometimes it is safer and more effective to evacuate to a tall building nearby (vertical evacuation), rather than moving to a distant refuge area. Check in advance whether there is some tall, sturdy building in your everyday environment that you can use in an emergency. In fact, even when a tsunami was near, many people have escaped and survived by climbing a palm or other tree in their garden or running to the second floor of their house.

At places further from the sea, a tsunami may come from almost any direction, and not necessarily from the sea. In particular, tsunamis move rapidly in rivers and other deep water, and can also approach by way of a river. In your everyday environment, there may be many places where the tsunami flow will become fast, or the water will suddenly become deep.

On the other hand, relatively safe high places are probably nearby. It is important to consider in advance how you will respond. For this, you must know the conditions around you (including the above information), and the predicted tsunami and the flows it will cause.

Being able to swim is also important, and regular training increases the probability of survival. In particular, it is extremely practical and effective to be trained in swimming with your clothes on. Today, there are various facilities where you can try swimming in a current and experience its difficulty under simulated conditions. For example, in swimming pools with wave generators, you can experience the force of breaking waves, and in pools with artificial currents, you can get a real feeling of how easily you can be swept away, and how difficult it is to swim against these currents. However, when people are swimming slowly, its speed is only at around 0.3 m/s, which is much slower than the speed of a tsunami flow. Thus, you must understand that being able to swim does not mean you will be safe.

References

Takahashi, S., Endo, K., and Muro, Z. (1992): "Research on People Falling on a Seawall during Overtopping," Report of the Port and Airport Research Institute (PARI), Vol. 31, No. 4, pp. 3-31. (in Japanese)

Arikawa, T. et al. (2006): "Large-scale Experiment on Run-up Tsunami Forces," Proceedings of Coastal Engineering, Japan Society of Civil Engineers (JSCE), Vol. 53, pp. 796-800. (in Japanese)

Takahashi, S. et al. (2008): A Study of Drowing Risk during Evacuation from Tsunami, Annual Jour. of Civil Engineering in the Ocean, Vol. 24, pp. 159-164. (in Japanese)

http://www.pref.ehime.jp/060nourinsuisan/080ringyou/00001461021016/3_rinsan/22.htm (in Japanese)

2.3 How Houses are Washed Away

A video was taken at Banda Aceh in Northern Sumatra during the Great Indian Ocean Tsunami of 2004, which shows the tsunami flowing like a great river in a flood. This video was taken from the second floor of a house (Fig. 2.12) about 2.3 km inland from the coast. The inundation

Fig. 2.12 A house which survived both the earthquake and tsunami.

depth at this house and the surrounding area was approximately 3.9 m. Such video recording was possible because the house was not destroyed by the earthquake or the tsunami. This fact shows that a sturdy building can also function as a tsunami refuge building. Knowing the conditions and factors in the destruction of buildings by a tsunami, and the conditions and circumstances under which buildings are washed away, is extremely useful for surviving a tsunami.

(1) *Conditions under which buildings are damaged*

The conditions for destruction of buildings differ, depending on the type of construction, structure, topography of the land behind the building, and other factors. Even with the same construction, the conditions for destruction are not always the same. This is because the quality of the materials used and the structure of the buildings differ, depending on the country and region. The method of joining columns and beams and the structure of walls are particularly important. For example, the strength of reinforced concrete (RC) buildings differs greatly, depending on whether the walls are of RC concrete or unreinforced brick. However, trying to explain the conditions for destruction based on a detailed classification of these factors would only complicate the matter, and would not be particularly meaningful. Similarly, due to the infinite diversity of damage, a detailed classification of the degree (condition) of damage would not necessarily be appropriate. Therefore, the following will use the three categories "heavy," "moderate," and "light" to classify the degree of damage. The definitions of these three terms are as follows.

Heavy damage: A significant part of the walls and columns are broken or lost. This category also includes complete destruction of both walls and columns. Restoration of the destroyed parts is not possible.

Moderate damage: Almost all columns remain, but the walls have partially been destroyed. Reinforcement or replacement of the columns is possible. With this degree of damage, restoration is possible.

Light damage: Windows and similar parts may be broken, but the walls are intact. This category includes cases where damage is limited to inundation by water and there is no mechanical damage. The building can be reused after minor repairs.

(1) *Relationship between inundation depth and degree of damage to buildings*

The relationship between inundation depth and the degree of damage was arranged by type of building construction based on examples of past damage (Fig. 2.13). In this figure, D (Destruction; □) is equivalent to the above-mentioned "heavy damage," PD (Partial damage; Δ) is equivalent to "moderate damage," and W (Water damage; ○) is equivalent to "light damage." Inundation depth is the depth on the side where the tsunami flow acts on the building. Here, the water level rises due to the damming effect of the building. (In the following, this side is simply called the "front" of the building.) The structure of the building, its distance from the sea, and the country and region are not considered.

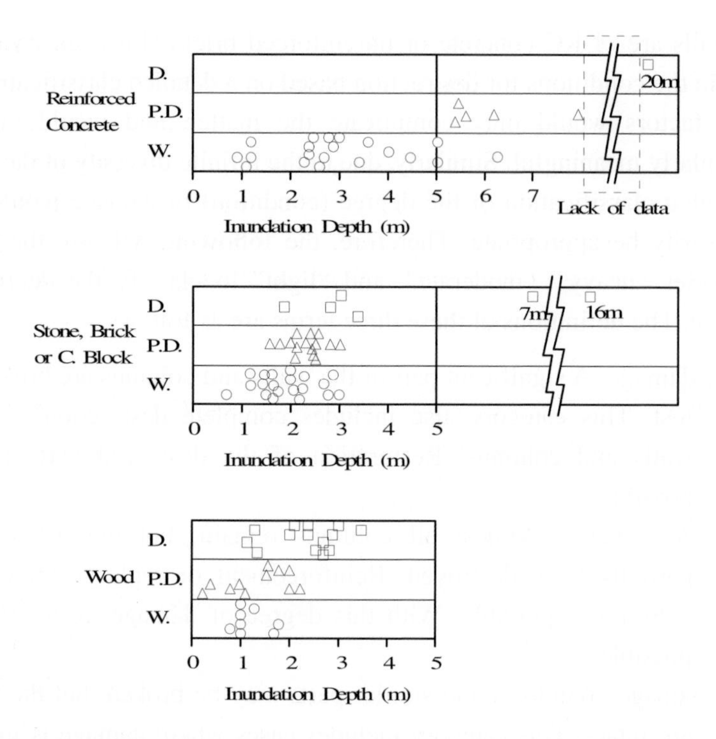

Fig. 2.13 Relationship between inundation depth and degree of damage classified by type of construction.

As shown in Fig. 2.13 (top), buildings of reinforced concrete construction show light damage when the inundation depth is less than 5 m and moderate damage when inundation is 5-8 m. Only one example of heavy damage is known, and this occurred with inundation of 20 m. The examples with moderate damage occurred during the Great Indian Ocean Tsunami in buildings constructed with unreinforced walls. In comparison with buildings with ordinary reinforced concrete walls, the strength of these buildings is slightly weak. Because no data are available for inundation depths from 8 m to 20 m, the critical condition for heavy damage is unknown.

In stone, brick, and block buildings, no clear difference can be seen in the degree of damage depending on the inundation depth as shown in Fig. 2.13 (middle). For example, light, moderate, and heavy damage can all occur at inundation depths of 2-3 m. According to a study on damage carried out prior to the Hokkaido Nansei-oki Earthquake Tsunami of 1993, damage was light when inundation was less than 3 m and heavy when inundation was more than 7 m. However, the relationship between inundation depth and damage becomes unclear when later examples are considered, such as the East Java Earthquake Tsunami of 1994 and the Indonesia/Biak Island Earthquake Tsunami of 1996. These additional examples of damage included heavy damage with an inundation depth of 2-3 m in a building with thin brick walls containing a small amount of thin wire. It is fair to say that the structure and other features of stone, brick, and block buildings differ depending on the country and region, and their strength is extremely diverse.

In wood-frame buildings, damage tends to increase from light to moderate and heavy as the inundation depth increases, but the boundaries for these levels of damage are not clear (Fig. 2.13, bottom). In examples up to 1993, damage had been light at inundation depths of less than 1.5 m, moderate at 1.5-2 m, and heavy at more than 2 m. However, when recent examples are considered, the relationship between inundation depth and damage becomes unclear. There is one example where moderate damage was caused at an inundation depth as small as 0.25 m. Its cause is thought to be impact by floating objects during the tsunami.

(2) *Relationship between flow velocity or drag force and degree of damage*

When the leading edge of a tsunami passes during run-up on land, a large amount of seawater surges through the area. Drag force acts on buildings as a result of this flow. Because the velocity of a tsunami on land is related to the inundation depth, slope of the water surface, slope of the ground, roughness of the ground surface, and other factors, it is extremely difficult to estimate its size. Estimation was attempted based on the inundation depth, h and the flow velocity, u. The flow velocity was estimated from the difference between the inundation depth at the front and rear side of a building (Fig. 2.14). In this figure, the solid lines

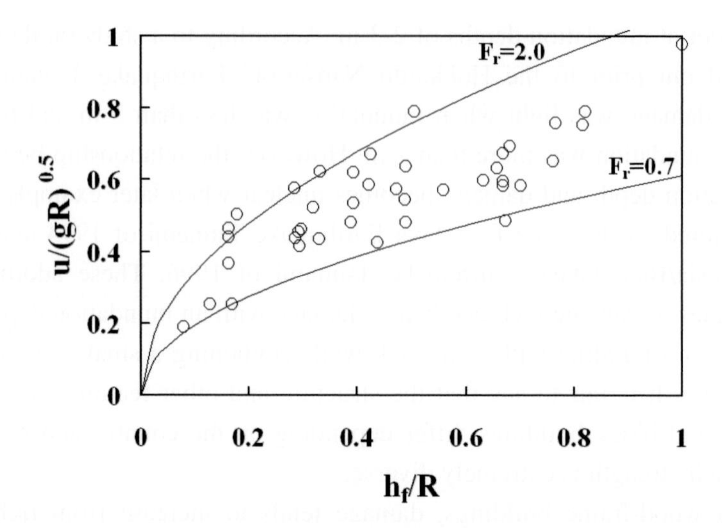

Fig. 2.14 Relationship between inundation depth (front side) and flood flow velocity.

are an envelope of all data using $F_r(=u/\sqrt{g})$ as a parameter, g and R denotes the acceleration of gravity, and the height of a tsunami at or near the measurement point using the sea surface as a standard, and the subscript f the front and rear side of a building, respectively.

In Fig. 2.14, the upper side of the envelope is the flow velocity at which the flow is more dangerous to buildings, etc. A calculation based on $F_r = 2.0$ gives the following equation.

$$u/\sqrt{gR} = 1.1\sqrt{h_f/R} \tag{3}$$

Deforming this equation,

$$u = 1.1\sqrt{gh_f} \tag{4}$$

Using this equation, the dangerous flow velocity can be calculated from the inundation depth at the front side of a building.

The force (drag force) F_D acting on buildings, etc. is assumed to be proportionate to the square of the flow velocity, and can be estimated by the following equation.

Table 2.2 Standards for judging degree of damage by type of construction.

Type of construction	Moderate damage			Heavy damage		
	h_f (m)	u (m/s)	F_D (kN/m)	h_f (m)	u (m/s)	F_D (kN/m)
Reinforced concrete	-	-	-	8.0 or more	9.7 or more	430~782 or more
Concrete block	3.0	6.0	60.5~110	7.0	9.1	329~598
Wood	1.5	4.2	15.1~27.5	2.0	4.9	26.9~48.9

$$F_D \cong 0.61 \gamma_w C_D {h_f}^2 B \tag{5}$$

where, γ_w is the weight of seawater per unit of volume, C_D is a drag force coefficient (1.1-2.0), and B is the breadth of a building in the flow direction.

Table 2.2 shows the results of an estimation of the flow velocity u and drag force F_D using the inundation depths, h_f (Fig. 2.14), corresponding to moderate and heavy damage of buildings. The critical condition for heavy damage of a wood-frame building is a flow velocity of 4.9 m/s. When calculated at the lower line of the envelope in Fig. 2.14, the flow velocity for heavy damage of wooden buildings is only 2.8 m/s. Accordingly, if information on a tsunami attack is received, you should not stay in a wooden building, but rather, should evacuate quickly to a safer place while escape is still easy, before the area is inundated.

(2) *Conditions for washing away of buildings, etc.*

(1) *Factors in washing away*

The main external forces which cause buildings to be washed away are the wave force of the tsunami, the hydraulic force of the flowing water, buoyancy, and uplift pressure. Impact against buildings by floating objects can also contribute to buildings to be washed away.

(a) *Wave force*

When a wave exerts a force on a structure, etc., the force per unit of area is called the wave pressure, and the total force acting on the structure is called the wave force. When a tsunami floods over flat land after overtopping a vertical seawall, its wave pressure is considered to have the hydrostatic pressure distribution. In contrast, the wave pressure at impact of the leading edge of a tsunami is thought to be very large. Recently, large-scale tsunami experiments have become possible, and researchers have begun to study the effect of the wave pressure when the leading edge of a tsunami flow flooding across land impacts against buildings. This is called the "surge front wave pressure." The results show that the vertical distribution of the surge front wave pressure differs in the status of breaking, bore, and overflow. When a wave breaks on the front side of a building, it is clear that an extremely large impact wave force is generated.

(b) *Hydraulic force of flow (drag force)*

After a tsunami propagates as a bore or wave-shaped bore on a seabed with a uniform slope, hydraulic force of flow (drag force) acts on buildings, etc. (Eq. (5)). Such hydraulic force acts during run-up of a tsunami on land, that is, when the tsunami flows landward from the ocean, and also during the return flow, when the sea water returns to the ocean. There are many examples of damage during the return flow. Damage during the return flow can be explained as follows.

Here, we consider the shape of the water surface of a tsunami on a uniformly-sloping ground along its cross-section in the land-to-ocean direction. The shape of the water surface during run-up becomes "convex." In other words, it has a shape that is swelled upward. Therefore, the thickness of the flowing water (water depth) is large, and the slope of the water surface is small. On the other hand, during the return flow, the shape of the water surface becomes "concave"; i.e., it has a shape that is sunken downward. Therefore, the thickness of the flowing water is small, and the slope of the water surface is large. As a result, when the sea water that flooded the land returns to the ocean, the flow velocity becomes large and tends to cause much damage.

In places where the topography of the land causes a concentrated flow of sea water during return flow, damage occurs during the return flow. Furthermore, in some cases, the structural strength of buildings against the return flow is reduced. In extreme examples, because coastal seawalls are constructed assuming attack by waves from the sea, they are almost completely defenseless against flows from the land.

(c) *Buoyancy*

Needless to say, structures on land, including buildings, are designed without considering buoyancy. Here, however, let us consider buildings on ground with high water permeability, houses with air vents below the floor to prevent dampness, and the like. When a tsunami floods the land, water may enter under the bottom of the building or the floor of the house, generating upward water pressure. As a result, buoyancy acts on the building. This tendency may be even more remarkable in recent houses with high air-tightness. Moreover, buildings with basements may receive of "double punch" of buoyancy from water flooding the basement as well as liquefaction of the ground.

In addition to buoyancy, uplift pressure is a cause of damage in buildings on ground with high water permeability and houses with air vents below the floor to prevent dampness. Uplift pressure is caused by dynamic pressure on the walls of the building, and acts on the bottom of buildings and underside of the floors of houses. The uplift pressure caused by the flow that is flooding the land can reach 10% to 30% of the buoyancy. This uplift pressure depends on various factors, such as the flow velocity, inundation depth, and the width-to-depth ratio of buildings, and is still not understood in detail.

(d) *Floating objects*

A flood flow carrying a large number of floating objects, such as the debris from wrecked houses, furniture, drum cans, etc. is shown in a video taken at Banda Aceh in Northern Sumatra during the 2004 Indian Ocean Tsunami (Fig. 2.15). When the flood flow caused by a tsunami is accompanied by a large amount of floating objects, the flow velocity and the velocity of the floating objects are reduced by friction between the floating objects and the ground and between the objects themselves. As the flow velocity decreases, the inundation depth tends to increase. Neither the decrease in velocity nor the increase in depth is understood quantitatively. The impact force of tsunami flood flows carrying floating objects is also unknown. These subjects will require investigation and research in the future.

Typical examples of floating objects are drifting boats and driftwood. Because both boats and wood naturally float in seawater, there is a high possibility that these objects will drift extremely long distances.

Among ships and boats, flat-bottomed boats have a shallow draft relative to their size and can drift far inland. As a result, they can easily cause enormous damage. Examples of these include a power generating barge at Banda Aceh in Northern Sumatra during the 2004 Indian Ocean Tsunami (Fig. 2.16) and a bulk cargo barge at Leupung on the western coast (Fig. 2.17). The power generating barge, which weighed 2,500 tons, drifted about 3 km inland. The bulk cargo barge destroyed buildings, flattening a coastal forest as well.

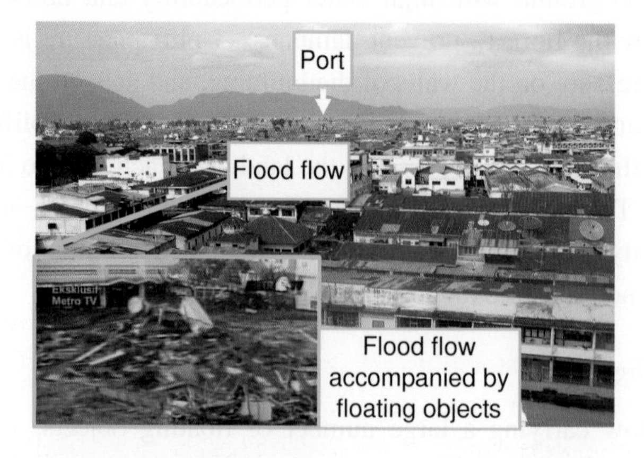

Fig. 2.15 Flood flow accompanied by floating objects.

Fig. 2.16 Damage to buildings by a drifting 2,500 ton power generating barge.

Fig. 2.17 Damage to buildings and a coastal forest by a drifting bulk cargo barge.

In Japan, many ports have coastal timber storage basins. During tsunami attacks, the stored timber is frequently floated out of the timber basin, increasing the damage. However, no timber basins have taken full-scale countermeasures to prevent timber outflow during tsunamis, and no effective outflow countermeasures law has been established. In contrast to this situation, progress has been made in research on the behavior of driftwood after outflow, and the impact force of driftwood can now be estimated quantitatively from the inundation depth and parameters related to the driftwood (diameter, length, weight, strength, etc.). Recently, study of the probability of simultaneous impact by multiple pieces of driftwood and other issues was also begun. If driftwood or other floating objects (such as fishing boats) become caught across two buildings, the shielded area of the flow and the water level will increase, and as a result, large drag force will act on the buildings. This force can become far greater than the drag force acting on the buildings when drifting objects are not involved, and result in destruction of the buildings.

Other floating objects include aquaculture rafts, fishing gear, the debris from wrecked buildings, furniture, and so on. In port areas, shipping containers, automobiles, and oil tanks also become floating objects.

(2) *Conditions and circumstances of washing away (wood-frame houses and oil tanks)*

When a building is washed away, reuse is generally impossible. Therefore, the degree of damage must be considered the same as "heavy damage," as described in 2.3(1). The conditions for washing away have been studied in examples collected up to the present. Because many examples are available for wooden buildings, the conditions for wooden buildings are understood to some extent. With wooden buildings, washing away occurs when the inundation depth on the front side of the building is 2 m or more. However, conditions for washing away differ in buildings anchored to a foundation (strip footing type foundation) and buildings not anchored to a foundation. Buildings which are not anchored to a foundation are easily washed away. In an example at Crescent, California (USA) from the 1964 Alaska Earthquake Tsunami, wood-frame houses which were not anchored to foundations were floated by buoyancy and damaged when the inundation depth was only 1.2 m.

Fig. 2.18 An oil tank that floated and was washed away.

Strip footing foundations are not buried deep in the ground. However, in many cases, the foundation is not washed away, as might be expected, and remains even after the upper part of the structure is washed away.

The problem of washing away is not limited to buildings. During the 2004 Indian Ocean Tsunami, in the port of Krueng Raya to the east of Banda Aceh in Northern Sumatra, three empty oil tanks (diameter: 17.1 m, height: 11.1 m) were washed away when its draft became 2.1 m (Fig. 2.18). This draft was estimated from leaked oil stains left on the side of the tank. The tanks were not anchored to foundations. It is thought that buoyancy on the order of 500 tons acted on the tanks, causing them to float.

(3) *Circumstances of washing away*

The circumstances of buildings which are washed away differ, depending on the mode of tsunami attack, topography of the land, distance from the ocean, whether the building is anchored to a foundation or not, and other factors.

When the leading edge of a tsunami forms a breaking wave and runs up on flat, low land, the flow velocity is large, and the changes of the water level over time are extremely large and rapid. Therefore, the governing forces acting on buildings are the wave force and drag force. Destruction of the building occurs first, regardless of whether the building is anchored to a foundation or not. This is not limited to wood-frame buildings. After destruction of the building, parts of the building are washed away together forming a mass of a certain size, while others are washed away in loose pieces. In some cases, the lower part of the building is destroyed, and the upper part is washed away with its existing shape still intact.

In cases where a cliff or other barrier is located directly behind a building, the horizontal flow velocity will be weak, and the water level will gradually rise. In such cases, buoyancy is the governing force acting on the building. If the building is not anchored to a foundation, it will float in its existing shape, whether it is wooden or RC.

In Fig. 2.19, the roof sections of destroyed buildings are being washed away. This scene was video-recorded from the second floor of the house in Banda Aceh in Fig. 2.12. Tin roofs are popular in this area. Such roofs are being washed away together with the wooden roof framing.

Fig. 2.19 Roof sections of destroyed buildings being washed away.

Fig. 2.20 Washed-away second floor of a wood-frame building (Aonae, Okushiri, Japan, 1993).

In Fig. 2.20, a partially second-storied house was washed away. The first floor was destroyed, and the second floor was washed away. The remains were stopped by a small rise in the ground. It is unclear whether the first floor was destroyed by wave force or was destroyed while drifting. However, in this example, a house anchored to a foundation was washed away by a tsunami that did not reach the second floor.

In Fig. 2.21, a reinforced concrete building was washed away. Again, the ground floor was destroyed, and the second floor was washed away and left blocking a road. Although the distance may be shorter than with wooden buildings, a large tsunami can wash away RC buildings with their shape intact. Unfortunately, the inundation depth at this building is unknown. Therefore, this example was not reflected in the conditions for destruction of buildings in section (1).

Fig. 2.21 Second floor of an RC building blocking a road after being washed away (Northern Sumatra, 2004).

References

Asakusa, R., Iwase, K., Ikeya, T., Takao, M., Kaneto, T., Fujii, N., and Omori, M. (2000): An Experimental Study on Wave Force Acting on On-Shore Structures due to Overflowing Tsunamis, Proceedings of Coastal Engineering, Japan Society of Civil Engineers (JSCE), Vol. 47, pp. 911-915. (in Japanese)

Arikawa, T., Ohtubo, D., Nakano, F., Shimosako, K., Takahashi, S., Imamura, F., and Matsutomi, H. (2006): Large Model Test on Running-up Tsunami Force, Proceedings of Coastal Engineering (JSCE), Vol. 53, pp. 796-800. (in Japanese)

Iizuka, H. and Matsutomi, H. (2000): Damage due to the Flooding Flow of Tsunami, Proceedings of Coastal Engineering (JSCE), Vol. 47, pp. 381-385. (in Japanese)

Ikeno, M. and Tanaka, H. (2003): Experimental Study on Impulse Force of Tsunami Running Up to Land and Drifting Objects, Proceedings of Coastal Engineering (JSCE), Vol. 50, pp. 721-725. (in Japanese)

Matsutomi, H. (1991): Pressure Distribution and Total Wave Force by Impact of Breaking-Wave Bore, Proceedings of Coastal Engineering (JSCE), Vol. 38, pp. 626-630. (in Japanese)

Matsutomi, H., Satonaka, Y., and Ikeda, H. (1993): Actual Condition of Coastal Timber Storage Basins, Tsunami Engineering Technical Report, No. 10, pp. 29-42. (in Japanese)

Matsutomi, H. and Shuto, N. (1994): Tsunami Inundation Depth, Flow Velocity, and Damage to Buildings, Proceedings of Coastal Engineering (JSCE), Vol. 41, pp. 246-250. (in Japanese)

Matsutomi, H. and Iizuka, H. (1998): Tsunami Current Velocity on Land and Its Simple Estimation Method, Proceedings of Coastal Engineering (JSCE), Vol. 45, pp. 361-365. (in Japanese)

Matsutomi, H. (1998): Impact Force of Driftwood Accompanying Tsunami Flooding on Land, Gekkan Kaiyo, Kaiyo Suppan, Extra Issue No. 15, pp. 153-158. (in Japanese)

Matsutomi, H., Koshimura, S., Takahashi, T., Moore, A., Imamura, F., Kawata, Y., and Matsuyama, M. (2000): The 1999 Vanuatu Tsunami - Its Features and Future Problems, Proceedings of Coastal Engineering (JSCE), Vol. 47, pp. 336-340. (in Japanese)

Matsutomi, H., Omukai, T., and Imai, K. (2004): Fluid Force of Tsunami Flood Flows on Structures, Annual Journal of Hydraulic Engineering, Vol. 48, pp. 559-564. (in Japanese)

Matsutomi, H. and Tanabe, J. (2006): Experiment on Horizontal Diffusion and Transfer Diffusion of Driftwood – Probability of Impact by Multiple Pieces of Driftwood, Proceedings of Coastal Engineering (JSCE), Vol. 53, pp. 186-190. (in Japanese)

Matsutomi, H., Fujii, M., and Yamaguchi, T. (2007): Fundamental Experiment and Modeling of Flood Flows Accompanied by Drifting Objects, Proceedings of Coastal Engineering (JSCE), Vol. 54 (submitted). (in Japanese)

Mizutani, N., Takagi, Y., Shiraishi, K., Miyajima, S., and Tomita, T. (2005): Study on Wave Force on a Container on Apron due to Tsunamis and Collision Force of Drifting Container, Proceedings of Coastal Engineering (JSCE), Vol. 52, pp. 741-745. (in Japanese)

Matsutomi, H., Sakakiyama, T., Nugroho, S., and Matsuyama, M. (2006): Aspects of Inundated Flow Due to the 2004 Indian Ocean Tsunami, CEJ, Vol. 48, No. 2, pp. 167-195.

Shuto, N. (1993): Tsunami Intensity and Disasters, Tsunamis in the World, Kluwer Academic Publishers, Dordrecht, pp. 197-216.

Shuto, N. (2003): Damages to Houses by the 1964 Tsunami in Crescent City, California, IUGG XXIII General Assembly, JSS07/09P/A02-003.

2.4 Effect of Natural Objects

Based on a survey conducted after the 2004 Great Indian Ocean Tsunami, some examples were reported that dense mangrove and evergreen forest belts reduced the damage to houses and villages behind them. Figure 2.22 shows the distribution of damage by the tsunami on a southern Indian coast. In the red areas, wood-frame houses were washed away and damage was heavy. The dark green areas are mangrove forest belts. Circles show the locations of villages. Heavy damage occurred in

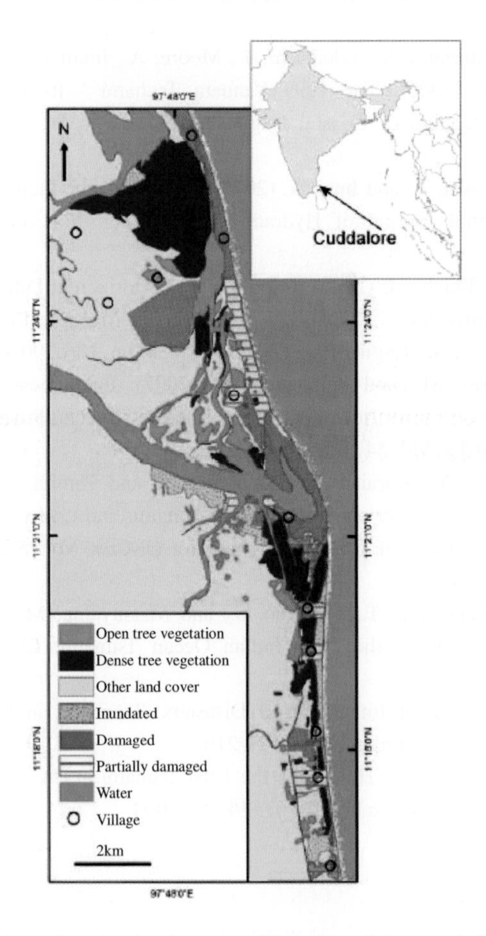

Fig. 2.22 Distribution of mangrove forests and location of damaged villages at Cuddalore in South India (Danielsen et al., 2005).

villages on the seaward side of the mangrove belts and on sandbars at the mouth of the river, where there were no mangroves, while villages located behind mangrove belts (on the landward side) were not damaged. That is, tsunami damage was reduced in villages behind or inside mangrove forest belts because the tsunami wave force was damped by the mangrove belts. Conversely, because dense forests do not grow on sandbars, the tsunami easily caused heavy damage. The disaster was even worse because there were no places to take shelter. Therefore, if possible, this type of sandbar should not be used as living areas. This section will explain the effect of stand-alone trees and forest belts consisting of clustered trees.

(1) *Effect of trees*

Stand-alone trees and sparsely-vegetated forest areas cannot be expected to reduce the force of a tsunami. However, they can save lives. The simplest example is a person climbing a tree to escape a tsunami (escape effect). Many examples exist. However, the following will introduce several from the Great Indian Ocean Tsunami. News videos broadcast in Indonesia and other countries shows people clinging to trees on the coast to avoid being swept away by the tsunami. It is also reported that students at Syiah Kuala University in Banda Aceh, Indonesia rode out the tsunami attack by clinging to tall coconut trees on the coast. At Patong Beach, Phuket Island in Thailand, people climbed trees and escaped to the second floor of a hotel. Thus, even a stand-alone tree can be useful in avoiding being swept away if you cling to it or escape to a high building by way of it.

However, the trees that grow along the coast tend to have shallow roots that spread near the ground surface. Young trees still have weak root systems and are easily toppled by a tsunami. Therefore, they are not suitable for shelter. In general, an adequate horizontal root system develops as a tree becomes older, and as a result, its ability to resist a tsunami increases. The trunk of a tree also becomes thicker as a tree ages. Although this increases the fluid force (drag force) in a tsunami, the fluid force is proportionate to the projected area of the trunk (in other words, the diameter), whereas the strength of the tree increases in proportion to its cross-sectional area of the trunk (square of the diameter). This means the strength of the tree against a tsunami increases as its diameter increases. In short, if you must cling to a tree or climb a tree, you must choose the thickest possible one. The height of a tree is generally said to be proportionate to the 3rd power of the diameter (d x 10^3), so choosing a tall tree is also effective. However, in areas close to the water, erosion of the coastal ground by the tsunami can cause trees to fall (Fig. 2.23(a) and (b)). Therefore, you should take shelter in a tree as far to the landward side as possible.

(a) Khao Lak Coast

(b) Pakarang Cape

Fig. 2.23 Trees toppled by erosion of the beach (southern Thailand).

In addition to the above, other effects of trees in reducing human injury include the “soft-landing effect,” in which people swept away by a tsunami are later washed up on soft forest belts, and the “trap effect,” in which flowing objects that could injure people (debris, scraps of lumber, etc.) are trapped by a forest.

(2) *Effect of greenbelts*

Here, forest belts near the coastal line are called “greenbelts.” Individual trees have the effect of reducing the flow velocity of a tsunami. The fluid force (drag force), which is proportionate to the square of the flow velocity, is also reduced as the flow velocity is reduced. Because trees in greenbelts grow in groups, the tsunami force reduction effect can be evaluated by the sum of the flow velocity reduction effects (and consequently, the drag force reduction effects) of the individual trees. Accordingly, the key parameter in an evaluation is the total projection area of the trunks of the trees in the inundation area which can resist the flow of the tsunami. Concretely, this means “(trunk diameter of trees x number of trees) x inundation depth.”

In order to imagine the tsunami force reduction effect of a greenbelt, a simple calculation was made for a case in which a 5-m tsunami attacks a greenbelt where the average trunk diameter of the trees is 40 cm. The fluid force (drag force) was estimated at the front (p_f) and rear (p_r) of the green belt, and the tsunami force (drag force) reduction effect was represented by the ratio C_R (= p_r/p_f). Figure 2.24 shows the relationship between the total number of trees per 10 m of coastal line and the tsunami force reduction effect obtained by calculation. The total number of trees is equal to “(width of greenbelt) x (density of trees in greenbelt) x 10 m of the coastal line.” For example, assuming the width of a greenbelt is 100 m, a total of 150 trees means 150 trees growing on land with an area of 1,000 m^2 (10 m in coastal direction x 100 m in landward direction). In this case, the density of the trees is 15 trees/100 m^2. Of course, with the same number of trees, the density of the trees will become higher as the width of the greenbelt becomes narrower.

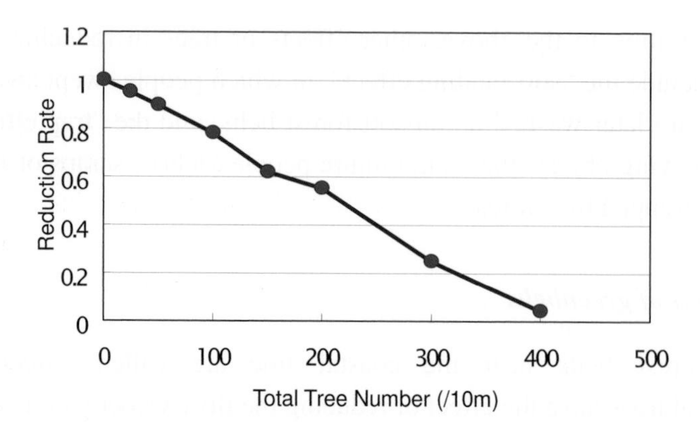

Fig. 2.24 Number of trees necessary per unit of coastal line (10 m).

According to this figure, the force of the tsunami is reduced to 0.8 with 100 trees per 10 m of coastal line; to less than 0.6 with 200 trees; and to nearly 0 with 400 trees. Thus, as the width of the greenbelt increases (with the same density of trees) or the density of trees increases, the power of tsunamis decreases and accordingly the degree of destruction or the impact during destruction become smaller. Therefore, effective tsunami countermeasures include building houses behind a greenbelt with a large total number of trees, and increasing the number of trees planted in front of existing houses. Constant care to preserve the greenbelt is also necessary. The effectiveness of greenbelts in reducing tsunami damage had already been reported before the Great Indian Ocean Tsunami, and planting of mangrove forests on some parts of the Indonesian coast had begun. Since the Great Indian Ocean Earthquake Tsunami, planting has been carried out at a faster pace. At present, Indonesia is carrying out a total of 18 tree-planting projects nationwide.

(3) *Latent risks in greenbelts*

In many cases, greenbelts contain roads for access to the coast. These roads are an unexpected latent risk. For example, during the Great Indian Ocean Tsunami, the tsunami attacked with tremendous speed along a 3-m wide road in a greenbelt on the coast at Khao Lak in Thailand. A local housewife stated that she was barely able to escape on this road on a motorbike. In other words, because the flow is attenuated in the greenbelt proper, the speed at which the tsunami approaches and its flow velocity increase along greenbelt roads, where there are no trees, and can make evacuation difficult during a tsunami attack. In this case, it is safer to run into the greenbelt and climb a tall tree, rather than fleeing straight to the inland area along the road.

A simple estimate of the tsunami flow velocity u_3 on the access road was calculated assuming the ratio of the width of an access road to the length of greenbelt in the coastal direction is an opening ratio γ, and the momentum of the tsunami is conserved before and after the greenbelt. An opening ratio of 0.1 corresponds to roads 10 m wide at intervals of 100 m in the coastal direction.

Figure 2.25 shows the ratio of the flow velocity in the access road u_3 to the tsunami flow velocity without a greenbelt u_1 for typical opening ratios, using the tsunami force reduction ratio C_R of the greenbelt. The curve is upward-sloping to the left with all opening ratios. In other words, as the tsunami force reduction ratio becomes smaller (i.e., as you go to the left side in the figure), the flow velocity on the road becomes faster.

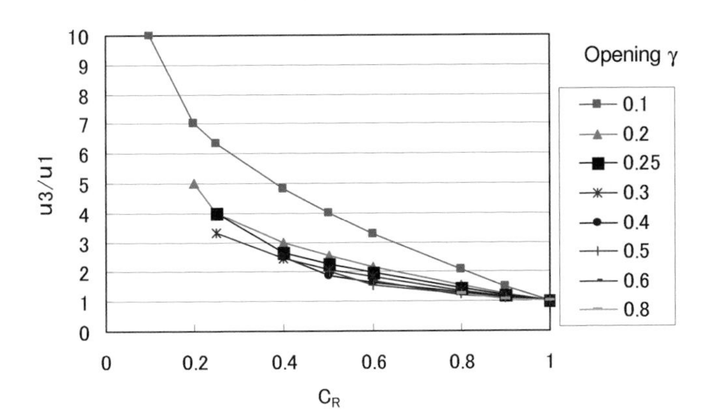

Fig. 2.25 Rate of flow velocity increase on roads in greenbelts.

As already mentioned, the tsunami force reduction ratio decreases (i.e., the tsunami force is reduced) as the width of the greenbelt becomes larger or the density of the trees increases. However, at the same time, the flow velocity on roads becomes faster, creating a dangerous situation. For example, when the opening ratio is small, at 0.1 (10-m wide road/100 m of coastal line), the tsunami is markedly concentrated on the road. If the tsunami force reduction ratio C_R is decreased to 0.2, the flow velocity on the road will be seven times faster than that without the greenbelt, and the danger during evacuation will be extremely high. Accordingly, in cases where the density of the greenbelt is high and its width is large (that is, small C_R), evacuation along the road should be avoided, and persons in danger should leave the road and take shelter in the greenbelt.

(4) *Issues related to greenbelts*

In the Great Indian Ocean Tsunami, there were examples in which the amount of coastal erosion was reduced by the existence of greenbelts. This is also an effect of greenbelts. However, if the strength of the ground is low or the scale of the tsunami is large, the greenbelt itself may be damaged. At Banda Aceh, Indonesia, the run-up height of the tsunami was more than 10 m, and the force of the tsunami was great. In addition, the ground was also weak in the town, which had developed on flat, sandy land at the mouth of a river. As a result of these factors, the greenbelt itself was washed away. Figure 2.26 shows the condition of a mangrove forest and tall trees after the Great Indian Ocean Tsunami. In (a), a sandbar at the coast has eroded, the roots of the trees are exposed to sea water, and the trees have begun to die. In (b), the sandbar itself was washed away, and sea water has flowed into the shallow land. As a result, the original mangrove forest area was lost. If this occurs, the greenbelt will have little effect in reducing the force of a tsunami. In the future, a comprehensive study to ensure the full effectiveness of greenbelts will be necessary, including study of the stability of the ground.

(a) Dying trees on a coastal sandbar

(b) Greenbelt area after the sandbar itself was washed away

Fig. 2.26 Condition of greenbelts in Banda Aceh, Indonesia.

References

Sasaki, Y., Homchuen, S., and Tanaka, N. (2005): The Great Indian Ocean Tsunami and the Role Played by Coastal Forests, – Coasts of Thailand and Sri Lanka –, Public International Symposium of the Japan Mangrove Society – Collected Papers on the 2004 Great Indian Ocean Tsunami and Mangrove Forests. (in Japanese)

Hiraishi, T., Arikawa, T., Minami, Y., and Tanaka, M. (2005): Field Survey on Indian Ocean Earthquake Tsunami Example Obtained Mainly in South Thailand, Report of the Port and Airport Research Institute (PARI), No. 1106, p. 20. (in Japanese)

Hiraishi, T., Minami, Y., and Tanaka, M. (2006): Hydrological Study of Reduction of Tsunami Force by Greenbelts, Report of the Port and Airport Research Institute, No. 1124, p. 17. (in Japanese)

Matsutomi, H. and Shuto, N. (1994): Tsunami Inundation Depth and Flow Velocity and Damage to Houses, Proceedings of Coastal Engineering, Japan Society of Civil Engineers (JSCE), Vol. 41, pp. 246-250. (in Japanese)

Matsutomi, H., Takahashi, T., Matsuyama, M., Harada, K., Hiraishi, T., Supartid, S., and Naksuksakul, S. (2005): The 2004 Off Sumatra Earthquake Tsunami and Damage at Khao Lak and Phuket Island in Thailand, Proceedings of Coastal Engineering (JSCE), Vol. 52, pp. 1356-1360. (in Japanese)

Thomas, P. (2005): Trees: Their Natural History (Japanese translation), Tsukiji Shokan, 4th ed., p. 149.

Danielsen, F., Sorensen, M.K., Olwig, M.F., Selvam, V., Parish, F., Burgess, N.D., Hiraishi, T., Karunagaran, V.M., Rasmussen, M.S., Hansen, L.B., Quarto, A., and Suryadiputra, N. (2005): The Asian Tsunami: A protective role for coastal vegetation, Science, Vol. 310, p. 643.

Dahdouh-Guebas, F., Jayatissa, L.P., Di Nitto, D., Bosire, J.O., Lo Seen, D., and Koedam, N. (2007): How effective were mangroves as a defense against the recent tsunami?, J. Current Biology, Vol. 15, No. 12, pp. 443-447.

Hiraishi, T. (2000): Characteristics of Aitape Tsunami in 1998 Papua New Guinea, Report of Port and Harbor Research Institute, Vol. 39, No. 4, pp. 3-23.

Hiraishi, T. and Harada, K. (2003): Greenbelt tsunami prevention in South-Pacific region, Report of the Port and Airport Research Institute, Vol. 42, No. 2, pp. 3-25.

2.5 Effect of Artificial Structures

(1) *Tsunami force reduction effect of breakwaters and their examples*

A breakwater is a structure which is constructed at the mouth of a harbor so as to reduce entry of ordinary waves from the open sea and thereby maintain a calm condition in the waters inside the harbor. Although breakwaters are essentially constructed as a measure against ordinary waves, they can also be expected to have various effects in protecting the harbor during a tsunami attack. Concretely, these effects include:

(1) Reducing the tsunami flowing into the harbor by narrowing the opening at the harbor mouth.
(2) Dissipating the energy of the tsunami by whirlpools generated when the tsunami passes through the opening in the breakwater.
(3) Avoiding resonance by the tsunami by changing the resonance period in the harbor.

These effects reduce the height and flow velocity of a tsunami in the harbor, and thereby reduce the inundation depth and inundated area in the port and its hinterland.

An example in which the damage due to a tsunami was reduced by a breakwater in the 2004 Great Indian Ocean Tsunami was confirmed at the Beruwela fishing port (Fig. 2.27) in northern Sri Lanka. Generally, the breakwaters at fishing ports are constructed so as to minimize waves in the port in order to ensure safe anchorage of the fishing boats. Similarly, at the Beruwela fishing port, an opening of approximately 90 m has been left to allow fishing boats to enter and leave the port, and a pair of breakwaters had been constructed across the remaining part to isolate the inner port from the open sea (Fig. 2.27(a)). The tsunami which attacked in this area ran up to a height of 4.82 m above sea level at the coast on the northern side of this fishing port (location indicated in Fig. 2.27(a)). Based on these facts, it is estimated that a tsunami of roughly 4 m or higher attacked the fishing port and the coast surrounding it. However, the height of the tsunami trace left on walls in the fishing port itself was 2.35 m above sea level, or more than 2 m lower than the height outside the port. This example demonstrates that breakwaters which are constructed for the purpose of obstructing the entry of ordinary waves into ports can also reduce invasion of ports by tsunamis. The inundation depth in the fishing port due to the weakened tsunami was 0.77 m. Although brick buildings suffered heavy damage in other areas,

(a) Bird's eye view of the fishing port (north: top of photo; taken on April 22, 2005)

(b) The village behind the fishing port, which largely survived the tsunami (photo taken on January 6, 2005)

Fig. 2.27 Example of reduced tsunami damage in Beruwela fishing port, Sri Lanka.

this kind of destruction did not occur at Beruwela. This is attributed to the fact that the destructive force of the tsunami was reduced due to the shallower inundation depth, together with the existence of a block wall between the port and the village. Although the block wall was destroyed by the tsunami, the village was largely spared due to the combined action of the breakwater and block wall (Fig. 2.27(b)).

(2) *Evacuation support effect of breakwaters*

Breakwaters can also be expected to have the effect of increasing the time that can be used for evacuation by delaying the arrival time of a tsunami, in other words, an "evacuation support effect." This will be explained using the results of a numerical simulation of a tsunami conducted for the simple V-shaped bay shown in Fig. 2.28(a). Figure 2.28(b) shows the time waveform of the tsunami at positions (a) through (d) in the plan view in the upper right part of Fig. 2.28(a). This simulation considers three cases, namely, a case in which there is no breakwater at the mouth of the bay, and cases in which breakwaters with openings of 400 m and 200 m are constructed. From the calculated results shown in Fig. 2.28(b), it is clear that the maximum tsunami height and the maximum inundation depth are both reduced in the waters behind the breakwater, that is, at point (b), and in the run-up areas, (c) and (d), when a breakwater exists, and this effect increases when the opening of the breakwater is narrowed. Focusing on run-up area (c), it is clear that, as the opening of the breakwater is reduced, the time to reach a certain inundation depth (e.g., 1 m) is delayed. This means that the time that can be used for an evacuation is increased, even in locations where the maximum inundation depth makes evacuation impossible, because the time for the inundation to reach that depth is delayed by the breakwater. In other words, breakwaters not only reduce the size and force of a tsunami during a tsunami attack, but also have an evacuation support effect.

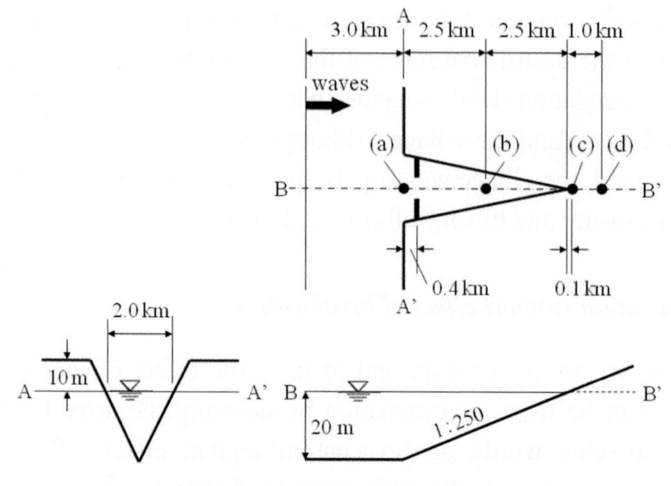

(a) V-shaped bay used in simulation

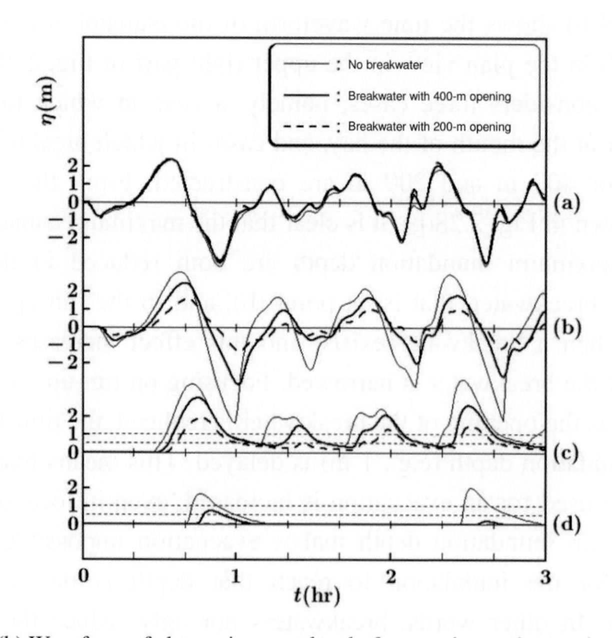

(b) Waveform of change in water level of tsunami at various points

Fig. 2.28 Tsunami simulation for model topography (Tomita, 2002).

(3) *Tsunami breakwaters*

In Japan, large-scale breakwaters have been constructed with tsunamis as their primary object, rather than ordinary waves, in order to positively utilize the tsunami force reduction effect of breakwaters. At Ofunato Bay, which suffered heavy damage in the 1960 Chilean Earthquake Tsunami, a large-scale breakwater was constructed at the mouth of the bay ("bay-mouth breakwater") in order to protect the bay as a whole from tsunami attack (Figs. 2.29 and 2.30). Construction began in May 1963 and was completed in March 1967. The following year, on May 16, 1968, the 1968 Tokachi-oki Earthquake Tsunami attacked Ofunato Bay. However, there was no inundation damage as a result of this tsunami, and the height of the tsunami measured at a tide station at the back of the bay was 1.2 m.

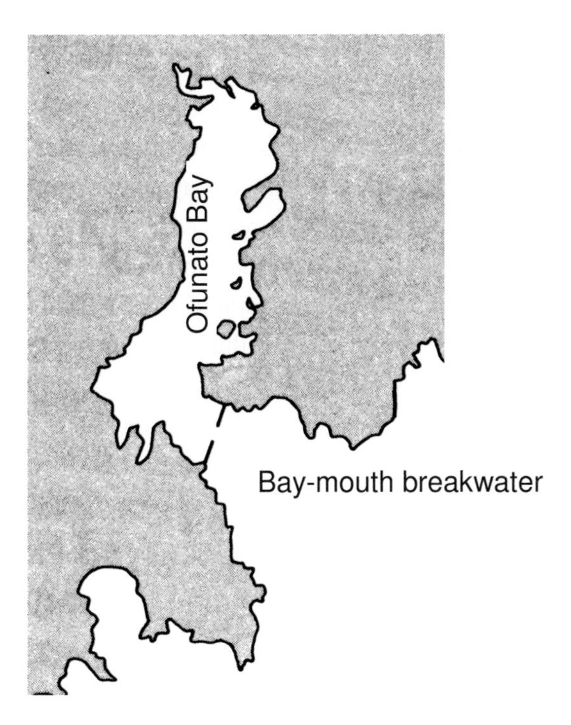

Fig. 2.29 Ofunato Bay, Japan and the location of the bay-mouth breakwater.

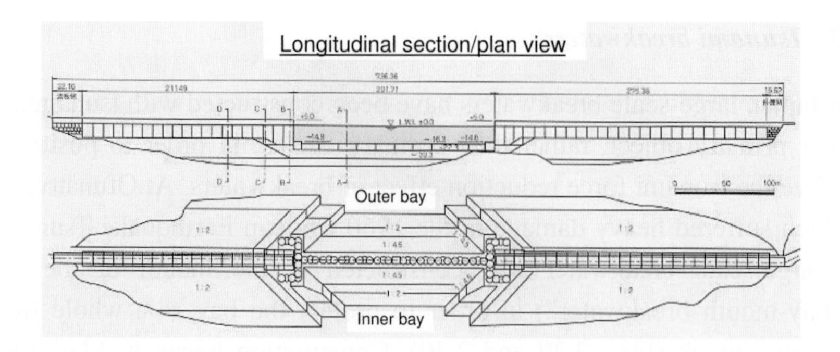

Fig. 2.30 Bay-mouth breakwater at Ofunato (source: homepage of the Ports and Harbors Section, Civil Works Dept., Iwate Prefecture).

When tsunami calculations were made using a numerical simulation model immediately after the tsunami attack, it was possible to reproduce the conditions that actually occurred, that is, no inundation and a tsunami height of 1.2 m at the back of the bay. Therefore, the effect of bay-mouth breakwaters was investigated by performing tsunami calculations under conditions with and without the bay-mouth breakwater using the same simulation model. Assuming no bay-mouth breakwater, the height of the tsunami that propagated to the back of the bay was 2.2 m. This means that the height of the tsunami was reduced by about half by the bay-mouth breakwater. In the calculations reproducing the tsunami, the speed of the flows (flow velocity) generated by the tsunami was also studied. This was not actually measured during the tsunami. The results showed a locally high flow velocity at the opening in the bay-mouth breakwater due the constriction of the flow when the tsunami passed through the narrow opening, but the flow velocity was slowed in the bay as a whole other than at this area. For example, when the flow velocity is 2 m/s without the bay-mouth breakwater, this is reduced to less than 1 m/s with the breakwater.

Because the function of a breakwater constructed at the mouth of a bay for the primary purpose of reducing the force of tsunamis was confirmed, this type of breakwater in particular is now called a "tsunami breakwater." When a necessary geographical condition (bay with narrow mouth) is given, this type of facility has an extremely high investment effect, because it is possible to protect the entire long coastal line inside a bay by concentrating the investment on the construction of the tsunami breakwater at the narrow bay mouth. On the other hand, if the inner bay is to be protected without constructing a tsunami breakwater at the bay mouth, it will be necessary to construct high embankments around the long coastal line. Needless to say, the total construction cost will increase as the total coastal line becomes longer, and it will also be necessary to sacrifice waterfront land with high land-use intensity, hindering social and economic activity in those areas. Moreover, in some cases, no land is available for the construction of large-scale embankments and similar facilities. These various problems can be avoided by constructing a tsunami breakwater. This is one important merit of this structure.

Table 2.3 shows the main tsunami breakwaters either completed or under construction in Japan. Among these, the tsunami breakwater at the Port of Kamaishi is being constructed in waters with a depth of 63 m, which is the world's deepest water site for a breakwater (Figs. 2.31 and

Table 2.3 Tsunami breakwaters major in Japan.

Location	Total length (m)	Opening (m)	Water depth (m)	Water area (km^2)	Construction period
Kuji Port (Iwate Pref.)	North: 2700 South: 1100	350	10-25	12	1990-2028 (scheduled completion)
Kamaishi Port (Iwate Pref.)	North: 990 South: 670	300	10-63	9	1978-2008 (scheduled completion) 2007 (virtually completed)
Ofunato Port (Iwate Pref.)	North: 244 South: 291	202	10-37	7	1963-1967
Susaki Port (Kochi Pref.)	East: 940 West: 480	230	2-18	2.9	1983-2007 (scheduled completion)

Fig. 2.31 Tsunami breakwater at Kamaishi Port (source: homepage a of Kamaishi Port Office, Ministry of Land, Infrastructure and Transport).

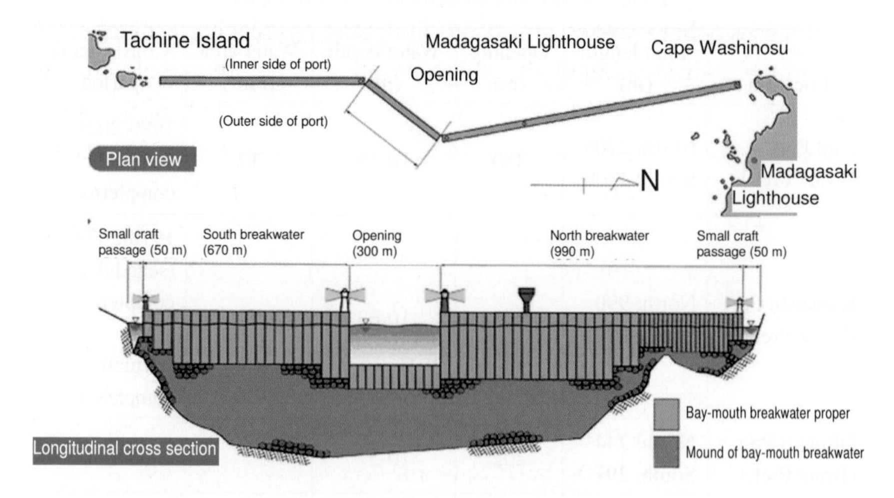

Fig. 2.32 Plan view and cross section of the Kamaishi Port tsunami breakwater (source: homepage b of Kamaishi Port Office).

2.32). The true size of this breakwater cannot be understood from the photograph, because the larger part of this structure is underwater. However, the caissons used in this tsunami protection breakwater are 30 m in length, 30 m in width, and 30 m in height, weighing 16,000 tons. When filled with sand, these caissons weigh 35,500 tons. The total volume of stone used in the mound under the caissons is 7 million m^3. This is equivalent to six times the volume of the Tokyo Dome baseball stadium, which can accommodate approximately 55,000 people.

(4) *Seawalls and embankments*

Seawalls and embankments are structures which prevent or reduce inundation of land by sea water resulting from tsunamis, storm surge, or waves (Figs. 2.33 and 2.34). In combination with this, they also have the function of reducing the force of waves by placing wave-dissipating blocks (called "wave-dissipating works" in the figure) or similar measures in their front. The difference between seawalls and embankments is that the ground height is raised by constructing a fill on land in embankments, but a fill is not required in seawalls.

In cases where waves are to be dissipated by placing a number of rows of wave-dissipating blocks, this measure is generally not effective if the width of the block structure is not 1/2 to 1/4 the length of the wave (wavelength). Because the wavelength of ordinary waves is on the order of several tens to 100 m, it is possible to construct effective and realistic wave-dissipating-type seawalls and embankments, and these have actually been constructed at many coasts. However, because the wavelength of tsunamis is from several 100 m to several kilometers, it is not possible to construct a wave-dissipating seawall of realistic scale which can dissipate the wave force of tsunamis. In cases where the wavelength of a tsunami is reduced by fission due to the effect of topography, it is not inconceivable that such structures may demonstrate a wave-dissipating function. In general, however, it is advisable to think that these structures will not have a function of diminishing the energy of tsunamis.

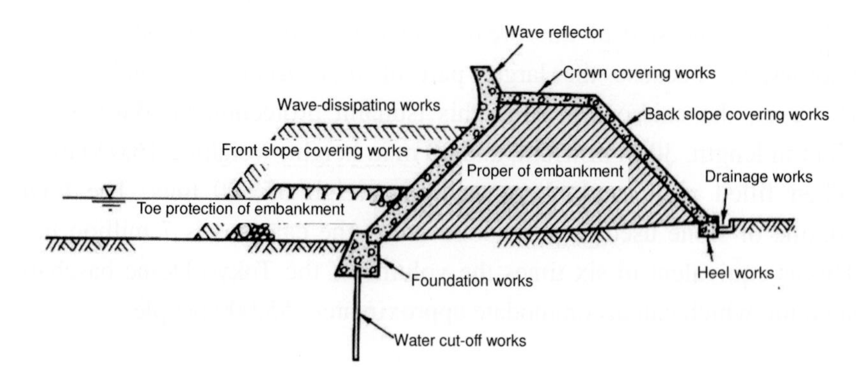

Fig. 2.33 Structure of an embankment (Japan Society of Civil Engineers; JSCE, 2000).

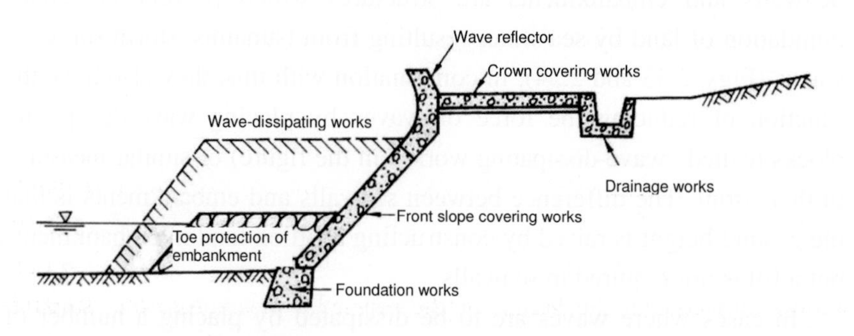

Fig. 2.34 Structure of a seawall (JSCE, 2000).

Essentially, seawalls and embankments are only effective against tsunamis when the crown height of the structure is higher than the height of the tsunami. The following will present three actual examples.

In the 1993 Hokkaido Nansei-oki Earthquake Tsunami, a run-up height of 11.7 m above sea level was recorded in the Hatsumatsumae District of Okushiri Island. Therefore, in order to protect villages in the coast in this area, a new embankment with a crown height of 11.7 m was constructed after the tsunami (Fig. 2.35). Because such a high embankment hinders everyday life, blocking sunlight and breezes, the land behind the embankment was raised, and residential land was created on this higher ground.

Fig. 2.35 Sea wall with a crown height of 11.7 m (Okushiri Island, Japan).

During the 2004 Great Indian Ocean Tsunami, inundation damage occurred at Malé Island, in which the capital of the Maldives (Figs. 2.36 and 2.37) is situated. The results of observation of the tsunami at a tide station showed that the height of the sea level just before the tsunami attack was approximately the same as mean sea level. The tsunami with a height of 1.45 m attacked there. The boundary of the inundated area was at rough 1.1 m above mean sea level. Speed bumps of about 10 cm high positioned across roads and other similar objects formed the boundary. This figure shows the eastern part of the island escaped inundation damage. The fact is that along the coast from the east of the island to the northeast which faces an outer atoll, a seawall with a crown height of 2.1 m above mean sea level had been constructed to prevent overtopping by high waves (Fig. 2.37). Inundation on land was prevented because the crown height of this seawall was higher than the height of the attacking tsunami. On the other hand, a large inundated area spread over the southern side of the island. A detached breakwater had been constructed off the southern coast (for details about the detached breakwater, see the following section). Because the detached breakwater was designed to

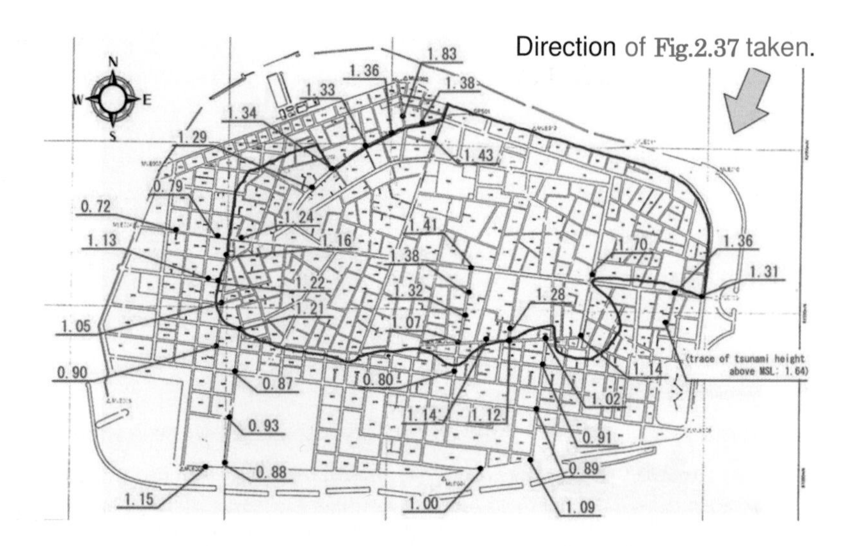

Fig. 2.36 Limit of inundation (blue line) and ground heights in Malé Island (red figures), the Maldives (Fujima et al., 2005).

Fig. 2.37 Appearance of Malé Island (photo taken on December 26, 2004; as for the direction of Fig. 2.37 taken, see Fig. 2.36; courtesy of Mr. Todd Rempel, Maldivian Air Taxi (Pte) Ltd.).

reduce the energy of ordinary waves, it was possible to reduce the height of the seawall behind it. In fact, at some locations, the crown height of the seawall was reduced by about 70 cm from that on the east side of the island. In other words, the level of protection against the tsunami by the seawall on the southern side of the island was lower. In addition, because the southern coast faced a channel between it and the neighboring atoll, there is a possibility that the tsunami was amplified as it passed through the narrow channel and a tsunami higher than in other areas attacked here.

Our third example is the Hiromura Bank in Hirogawa Town, Wakayama Prefecture, Japan, where the entire village was heavily damaged by inundation in the 1854 Ansei Nankai Earthquake Tsunami. The height of the tsunami is said to have been 5 meters. This area had also suffered tsunami attacks in 1605 and 1707. After the tsunami of 1854, a massive embankment was constructed in preparation against repeated tsunami attacks in the future. The project was also carried out to give employment to villagers who suffered losses in the disaster. The constructed embankment is 5 m high, 600 m long, and 2 m wide at the crown, and 20 m wide at the base (Fig. 2.38 and 2.39(b)). The area was

Fig. 2.38 Hiromura Bank (Hirogawa Town, Wakayama Pref., Japan).

(a) Inundated area in 1854 Ansei Nankai Earthquake

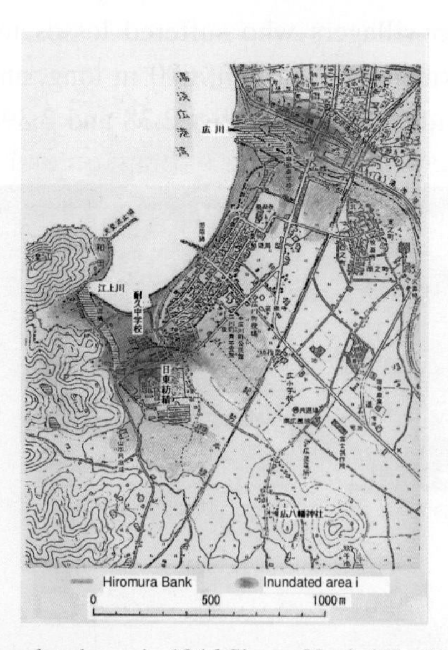

(b) Inundated area in 1946 Showa Nankai Earthquake

Fig. 2.39 Damage caused by tsunamis in Hiromura (source: homepage of Japan Meteorological Agency).

attacked by yet another tsunami 92 years later, in the 1946 Showa Tonankai Earthquake Tsunami. The height of this tsunami was 4-5 m. The Hiromura Bank, which had been constructed almost 100 years earlier, protected most of the residential area from inundation by the tsunami. In Fig. 2.39(a), the blue color shows the inundated area of Hiromura Village in the 1854 Ansei Tokai Earthquake; Fig. 2.39(b) shows the inundated area in the 1946 Showa Tonankai Earthquake. The hinterland behind the Hiromura Bank (red line in Fig. 2.36(b)) was not inundated.

(5) *Detached breakwaters and tsunamis*

Detached breakwaters are structures in which wave-dissipating blocks are piled up to above sea level offshore from the coast, basically parallel to the coastal line. Figure 2.40 shows a detached breakwater at the Kaike coast in Tottori Prefecture, Japan. This structure is termed a "submerged breakwater" when its crown is underwater; in particular, structures with wider widths are called "wide submerged breakwaters" or "artificial reefs." The functions of detached breakwaters are to reduce the force of waves attacking from the open sea, and to protect and enhance the sandy

Fig. 2.40 Example of a detached breakwater at Kaike, Tottori Pref., Japan (source: homepage of Hinogawa River Office, MLIT).

coast behind the breakwater by controlling sand drifts through reduction and diffraction of waves. The wave-dissipating function is realized by dispersing the wave energy by friction and formation of vortexes when sea water passes through the narrow space between the blocks.

Like wave-dissipating breakwaters, in order for a detached breakwater to effectively demonstrate a wave-dissipating function, it must have an adequate width relative to the wavelength of the object waves. Detached breakwaters which are effective against ordinary waves have been constructed. However, it is not possible to construct a detached breakwater of a realistic scale which is capable of dissipating the wave force of tsunamis, which have wavelengths of several 100 m to several kilometers. In addition, generally the crown of detached breakwaters is not high. Considering these facts, it can be said that detached breakwaters do not have a function of protecting the area behind the structure from tsunamis.

References

Itoh, Y., Tanimoto, K., and Kihara, I. (1968): Simulations of Effect of Breakwaters on Long Period Waves (4th Ed.), Effect of Ofunato Tsunami Protection Breakwater in 1968 Tokachi Earthquake Tsunami, Report of Port and Harbor Research Institute, Vol. 7, No. 4, pp. 55-83. (in Japanese)

Homepage of the Ports and Harbors Section, Civil Works Dept., Iwate Prefecture; Port of Ofunato Tsunami Protection Breakwater:
http://www.pref.iwate.jp/~hp0606/harbor/harborinfo/index.html (in Japanese)

Homepage a of Kamaishi Port Office: Port of Kamaishi (outline):
http://www.pa.thr.mlit.go.jp/kamaishi/port/km01.html (in Japanese)

Homepage b of Kamaishi Port Office: Planing of the bay-mouth breakwater:
http://www.pa.thr.mlit.go.jp/kamaishi/sdy/minato/m0104.html (in Japanese)

Homepage of the Japan Meteorological Agency: "*Inamura no hi*" (Fire on harvested rice plant stacks)
http://www.seisvol.kishou.go.jp/eq/inamura/p7.html (in Japanese)

Goto, T. and Sato, K. (1993): Development of Tsunami Simulation System for Sanriku Coast, Report of Port and Harbor Research Institute, Vol. 32, No. 2, pp. 3-44. (in Japanese)

Shuto, N. (1992): Tsunami Intensity and Damage, Tsunami Engineering Technical Report, Disaster Control Research Center, Tohoku University, No. 9, pp. 101-136. (in Japanese)

Japan Society of Civil Engineers (2000): Design Manual for Coastal Facilities, p. 408, p. 582. (in Japanese)

Tomita, T. (2002): Present and Future of Storm Surge and Tsunami Countermeasures, Lecture Notes of the 2002 (38th) Summer Seminar on Hydraulic Engineering, B-8, p. 20. (in Japanese)

Tomita, K., Honda, K., Sugano, T., and Arikawa, T. (2005): Field Investigation on Damages due to 2004 Indian Ocean Tsunami in Sri Lanka, Maldives and Indonesia with Tsunami Simulation, Technical Note of the Port and Airport Research Institute, No. 1110, p. 36. (in Japanese)

Japan Port and Harbor Association (1999): Technical Standard for Port and Harbor Facilities and its Commentary, p. 594. (in Japanese)

Homepage of Hinogawa River Office: Coastal Projects: http://www.cgr.mlit.go.jp/hinogawa/work/kaigan.htm (in Japanese)

Hokkaido Regional Development Bureau: Hokkaido Development Club, Vol. 35: http://www.hkd.mlit.go.jp/topics/info/ippan/koho/graph/vol35/35_2.html (in Japanese)

Fujima, K., Tomita, T., Honda, K., Shigihara, Y., Nobuoka, H., Hanazawa, M., Fujii, H., Ohtani, H., Orishimo, S., Tatsumi, M., and Koshimura, S. (2005): Preliminary Report on the Survey Results of 26/12/2004 Indian Ocean Tsunami in the Maldives: http://www.nda/ac.jp/~fujima/maldives-pdf/ (in Japanese)

Chapter 3

Prevention and Mitigation of Tsunami Disasters

3.1 Viewpoint of Disaster Prevention (Preparedness)

Tsunamis generally occur only once in several decades to several hundred years. As a result, even persons who have experienced a tsunami tend to forget the wisdom and knowledge necessary to avoid damage. The importance of disaster prevention, that is, preparedness, was confirmed anew in the Great Indian Ocean Tsunami in December 2004. People saw the horror and tragedy of a tsunami in real images in their own living rooms. Many of the world's people understood the terror of tsunamis quite well. The video images taken during the tsunami attack were so powerful.

Japan is a tsunami-prone region where tsunamis have attacked repeatedly throughout history. Nevertheless, because tsunami disasters do not occur frequently, the necessity of fundamental countermeasures was not argued persistently and strongly enough, and the idea of steadily realizing plans from a long-term perspective is not necessarily adequate. Here is one problem: some people say that a major disaster like the Great Indian Ocean Tsunami will not occur in Japan. This is simply untrue. Even assuming that high-level countermeasures by hardware are in place, people must understand that a disaster on a huge scale is possible if they do not evacuate when necessary. Maintaining the confidence and awareness that a gigantic disaster shall not be allowed to occur is important for tsunami preparedness. Here, we will discuss the viewpoint of preparedness based on the conditions described above.

(1) *Creating a "disaster culture" in connection with tsunamis*

In order to avoid damage, an accurate knowledge of tsunamis is indispensable. It is also important that this knowledge become part of one's everyday habits (lifestyle culture) because only such knowledge embedded in everyday habits leads to prompt evacuation in case of emergency. "Lifestyle culture" in connection with disasters is called "disaster culture." The following are examples of such disaster culture related to tsunamis.

(1) A tsunami does not always begin with a "backwash":

In the Great Indian Ocean Tsunami, the 1st wave of the tsunami attack at Phuket in Thailand began with a backwash, where sea water retreated far offshore. However, in some cases, tsunamis begin with an incoming "rising wave." A person who knows only about one of these facts will delay evacuation when the tsunami begins with the other mode. To avoid death or injury, you must make a judgment based on signs at an earlier stage, and not wait until you can see changes in the water surface. If you feel an earthquake, stay away from the sea, and if you notice something strange at the beach, flee immediately. If you can do this, you are a person of disaster culture.

(2) If earthquake tremors continue for a minute or more, a tsunami will come:

Because the tremors were small in the 2004 Kii Hanto-oki Earthquake (Japan), many coastal residents judged that the tsunami would also be small. This is a terrible misunderstanding. You should fully understand that an earthquake with small tremors sometimes generates a large tsunami (in particular, this kind of earthquake is called a "tsunami earthquake"). This misunderstanding was also the reason why 22,000 people died in the 1896 Meiji Sanriku Earthquake Tsunami (Japan). In that earthquake, the intensity of the tremors was only 2 to 3. However, in spite of the fact that 20-30 minutes passed before the tsunami arrived, many people did not evacuate. The residents at the coast, who judged that the tsunami would be small, did not know this historical fact. Without this kind of knowledge, disaster culture is not possible.

(3) When tsunamis occur on the Pacific coast, large tsunami waves may continue to come one after another for six hours:

When the magnitude of an earthquake reaches as much as 8, a tsunami will attack in repeated waves. In large tsunamis, you must consider that the tsunami attack may continue as long as six hours. Fishing boats must not hurry back to port, and residents who have fled to refuge areas must remain there for six hours. If you are able to act based on this knowledge and experience, you have learned disaster culture.

(4) The "return flow" of a tsunami is a serious threat to villages on land that slopes downward to the sea:

The sea water that has run up on land during a tsunami attack accelerates as it returns to the sea, and can pull residents into the sea, along with debris from destroyed buildings and other objects. When this happens, it is difficult even to search for the dead. In the 1933 Showa Sanriku Tsunami, a total of 3,000 persons fell victims as dead or missing; out of these, 1,500 bodies were later recovered, but the remainder were simply lost in the sea. A tsunami attack on land is frightening, but it is necessary to recognize that the effects of the returning tsunami can also be terrible.

Without such specific knowledge of tsunamis, you may likely fall victim once you encounter an actual tsunami. It is necessary to nurture a disaster culture which makes it possible to translate this knowledge into action and thereby avoid damage. Without this disaster culture, serious damage is always possible.

(2) *Viewpoint of preparedness*

Disaster culture means that "action based on correct knowledge" becomes a lifestyle habit. There are three points that one must learn for this. In the case of a tsunami, as shown in Fig. 3.1, the first point is to "Know the mechanism of tsunamis." In learning the mechanism of

· To learn the mechanism of tsunamis
(Understanding of the mechanism of generation, propagation, and deformation of tsunami waves and the process of flooding by tsunamis)
· To learn the weaknesses to a tsunami
(Assumptions about human damage and physical damage to structures, buildings, etc.)
· To learn tsunami countermeasures
(Countermeasures based on knowledge and information comprise total disaster mitigation countermeasures, including management of countermeasures by structures and other "hard" facilities.)

Fig. 3.1 Three points to "Learn" in order to produce a disaster culture.

tsunamis as a physical phenomenon, the best method is to study as part of disaster preparedness education during science classes in school. The second point is to "Know the weaknesses in a tsunami disaster." These weaknesses include the weaknesses of structures and buildings to tsunami damage and the weaknesses in society, for example, surrounding people who require special protection during a disaster. In other words, this second point is not limited to visible strengths and weaknesses, like those seen in buildings. Historical examples teach that victims tend to be concentrated in certain particularly vulnerable segments of the population, namely, older persons, women, and children. The third point is "Know tsunami countermeasures." The actual content of countermeasures will be described in more detail later, but the basis is the information to be transmitted from the local and national governments.

In order to translate the knowledge learned in this manner into action, the process shown in Fig. 3.2 is essential. The process consists of the steps of "Learning," "Drilling," and "Exercising." In the "Learning" step, knowledge is learned by study. This is followed by "Drilling" that reflect the knowledge learned, and then by "Exercising" in which participants actually try performing actions. It is basically important to begin this process with the Learning step. It has been found that when local

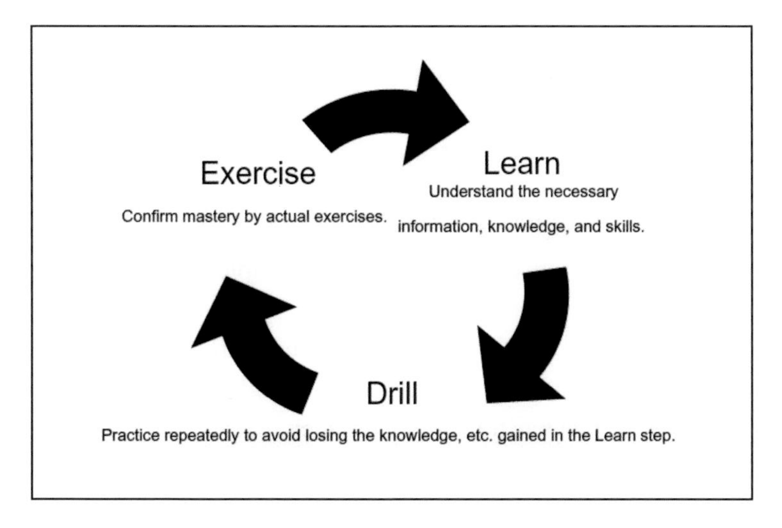

Fig. 3.2 Process of translating knowledge into action.

governments carry out evacuation drills or map exercises before residents properly learn about tsunamis as a physical phenomenon, such drills or exercises have the opposite effect, and the residents quickly forget their awareness of disaster preparedness. This process of translating knowledge into action is only completed when participants are given education in school or other training, their complete understanding is checked, and they are well prepared to use the knowledge when they actually encounter a tsunami.

(3) *Tsunami disaster mitigation system*

Countermeasures to prevent tsunami damage require a huge investment of money and time. But, no matter how great these efforts are, it is not necessarily possible to reduce damage to zero in a large tsunami. Generally, the words "disaster prevention" seem to imply some measures that reduce damage to zero. It is more realistic to accept the fact that some damage may be unavoidable, and aim at reducing that damage as far as possible. This is the concept of "disaster mitigation." Even with the concept of disaster mitigation, if the scale of the attacking tsunami is small, it is possible to reduce damage to zero; in other words, damage is prevented and disaster prevention ("zero damage") is achieved. In this sense, "disaster prevention" is a special case of "disaster mitigation." What methods can be used to reduce human and physical damage? No one countermeasure is a perfect solution. It is necessary to reduce damage as far as possible by reflecting local conditions to the extent possible and using a combination of multiple disaster mitigation measures.

When individual disaster mitigation measures are implemented separately, their effectiveness is limited. Therefore, it is necessary to develop and implement a strategic plan which realizes a synergistic effect by combining multiple measures. What is needed is a "strategic disaster mitigation plan." In the conventional thinking, there is a tendency to consider disaster prevention facilities ("hardware") and disaster prevention systems ("software") separately. However, in establishing a strategic disaster mitigation plan, it is necessary to consider hardware and software as an integrated whole. For example, a tsunami water gate is a disaster prevention facility (hardware), but by itself, it has no function. It only has a function when considered in combination with a disaster prevention system (software) such as under what conditions the gate is to be closed and how maintenance control is to be performed. The new concept that "hardware disaster prevention facilities are only one part of a software social disaster prevention system" is a development of this line of thinking. This concept is particularly important for tsunami preparedness.

When a strategic disaster mitigation plan is actually established, it is necessary to study which countermeasures shall be the object of concentrated investment. If possible, it is desirable that this study as well as the final decision-making process should be led by the local residents who may suffer damage in a tsunami. This is because the local residents will benefit from the results of their choices, and at the same time, they will also bear the risks and disadvantages of those choices. This embodies the principles of "self-help" and "self-responsibility." For this, however, the government must provide and make available all possible information. On the part of residents, effort in obtaining and understanding such information is important.

In developing disaster mitigation measures, first, it is necessary to prepare a detailed and concrete mid- to long-term implementation plan. If possible, mid- to long-term plans should be established at local level in the future, rather than at national or prefectural government level. Here, the term "local level" does not necessarily mean that the plan is to be prepared by city, town, or village units. It is more desirable to define such units from the viewpoint of disaster mitigation. This is because the range that is affected by certain disaster mitigation measures is best understood by the residents of the areas concerned. The aim of disaster mitigation measures is to reduce or avoid risk. Thus, assuming the perception of risk and the response to it differ depending on the individual, the effectiveness of disaster mitigation measures is ultimately measured by the sense of values of residents in the area affected by that risk. Based on the concept that the disaster prevention objectives and concrete disaster mitigation measures are to be established in each local area, the roles of the national and prefectural governments shall perhaps be limited to coordination from the viewpoint of the consistency of measures as a whole and implementation of large-scale disaster mitigation measures for wider areas. Thus, judgments as to the proper form of disaster prevention in respective areas shall basically be left to the discretion of each area. In this sense as well, the principles of self-help and self-responsibility of area residents are important as the basis of disaster mitigation.

(4) *Strategic disaster mitigation plans for tsunamis*

Figure 3.3 shows the composition of mid- to long-term strategic disaster mitigation plans. Plans consist of a Goal, Objectives, Targets (numerical targets with quantitative values and set periods), Policies, and Actions. A "Goal" is to achieve an unchanging vision established by the national, prefectural, or local government. An example of such visions is to reduce human damage to zero in case an earthquake occurs in the Nankai area.

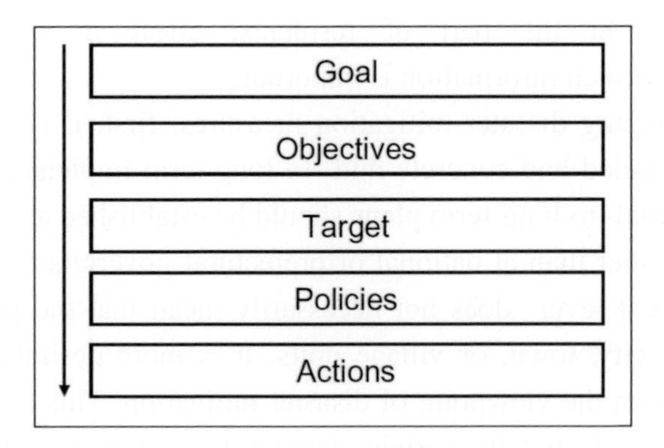

Fig. 3.3 Composition of mid- to long-term strategic disaster mitigation plans.

Realistically, however, it is difficult to achieve zero human damage when a tsunami more than 5 m in height strikes less than 10 minutes after an earthquake. Therefore, the local government sets "Objectives and Targets" to be achieved in several to five years from the present toward realizing this vision. Recently, the Japanese government proposed "Objectives and Targets" of reducing human damage due to a "direct-hit" earthquake under Tokyo or a Tokai, Tonankai, or Nankai earthquake by one-half in 10 years. (A "direct-hit" earthquake means an earthquake in which the epicenter is directly below a particular area.) These are "Objectives" with "Targets" expressed by concrete numbers. The measures necessary in order to achieve these are "Policies." The "Special Measures Law on Large-scale Earthquakes" for Tokai earthquakes and the "Special Measures Law for Promotion of Earthquake Disaster Prevention for Tonankai/Nankai Earthquakes" enacted by the Japanese government correspond to these policies. "Actions" realize the countermeasures set in these policies.

In order to determine the objectives in a tsunami disaster mitigation plan, it is necessary to make an assessment, which is broadly divided into three parts, as shown in Fig. 3.4. Needless to say, the content of the assessment will vary depending on the area. First, the local government must make an assessment to determine the extent of weaknesses (in other words, negative resources). This is called a "Resources assessment." It

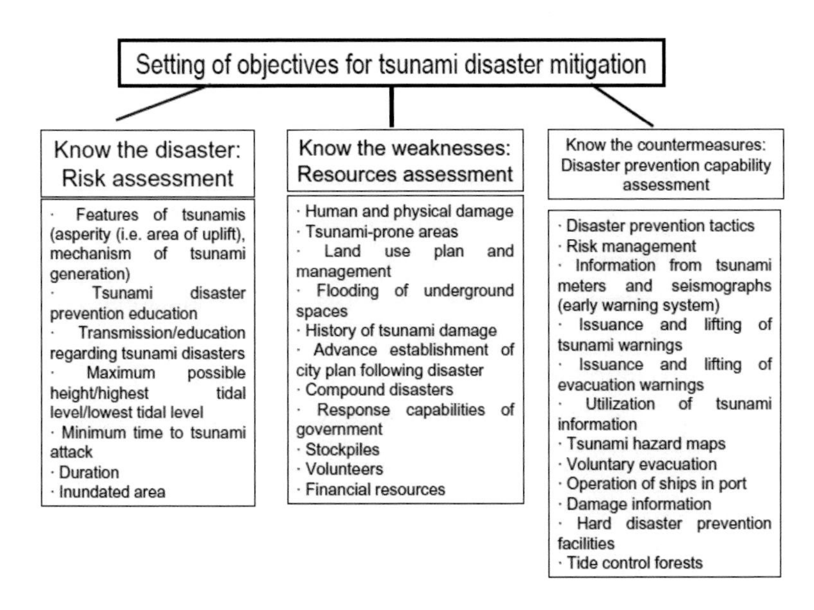

Fig. 3.4 Assessments for setting objectives in tsunami disaster mitigation plans.

is necessary to grasp the history of tsunami disasters in the area as one of the factors that compose the lifestyle culture as well as land use planning in the area and the possibility of occurrence of compound disasters, etc. It is also necessary to understand the levels of the damage evaluation and administrative capabilities of the local government and determine its risk management capabilities. This includes all items related to people, physical property, information, and financial resources. Quantitative understanding is necessary on the potential risks of tsunami disasters that exist in the area, the probability of their occurrence, and the expected size of external forces. This is called "Risk assessment." Simultaneously with this, it is also necessary to study the concrete content of countermeasures, disaster mitigation plan implementation capabilities, and its effectiveness. This is a "Disaster prevention capability assessment." If the evaluation in this study is not correct, it will not be possible to realize effective disaster mitigation countermeasures in advance. Based on these three assessments, objectives and targets are determined, policies are proposed, and an action plan is drawn up and

· Internal environment: SW
- What are the features of our area?
- S: Strength (What are its strengths?)
- W: Weakness (What are its weaknesses?)
· External environment: OT
- What are the features of conditions surrounding our area?
O: Opportunity (What are the positive factors?)
T: Threat (What are the negative factors?)

Fig. 3.5 SWOT analysis of tsunami disaster mitigation.

implemented. After the plan is implemented, the actual degree of its implementation needs to be evaluated periodically, and if implementation is inadequate, the process must be repeated from the three assessments. An example of this type of feedback loop is the "SWOT analysis" shown in Fig. 3.5.

(5) *Concrete example of strategic disaster mitigation plans*

In areas where adequate time is available for evacuation before a tsunami attack, the objectives may be the same as the vision or goal of the plan. In other words, the objective may be set at reducing the deaths due to a tsunami to zero. Figure 3.6 shows an example in which Osaka Prefecture established this kind of strategic plan. What is worth noting here is the fact that it is necessary to obtain the agreement of the local residents on these policy items. For this, complete information must be provided from the governmental side. The content of the action plan becomes more specific through this process.

Figure 3.7 is an example of a strategic plan when the objective is to achieve a speedy response to a wide-area tsunami disaster like that expected in a Tokai, Tonankai, or Nankai earthquake. As can be

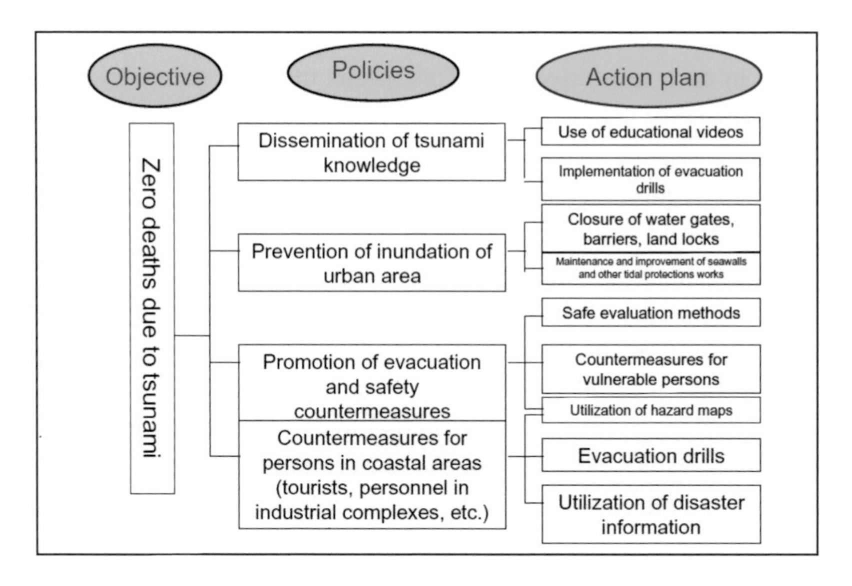

Fig. 3.6 Tsunami disaster mitigation plan by Osaka Prefecture.

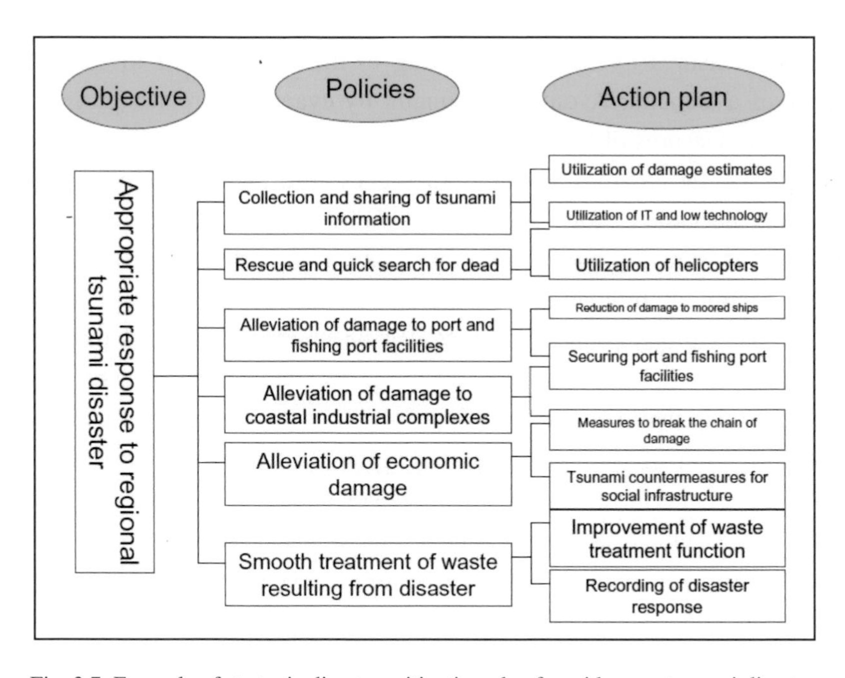

Fig. 3.7 Example of strategic disaster mitigation plan for wide-area tsunami disaster.

understood from this, it is necessary to consider what shall be adopted as policies in each respective area. This means that it is necessary to hold workshops for large numbers of related persons, and to build a consensus regarding the nature of the problems in that area. The order of priority is determined by this procedure.

3.2 Countermeasures by Hardware

The main purpose of tsunami countermeasures is to mitigate human damage (death and injury) and physical damage. The human damage and physical damage occur when the external force of a tsunami exceeds the limits that persons, objects, and the society can withstand. Therefore, tsunami countermeasures can be broadly divided into measures that reduce external force as far as possible, measures that increase the limits which can be withstood, and escape from the tsunami itself. From this viewpoint, the construction and reinforcement of tidal protection works such as breakwaters, seawalls, and other structures can be regarded as countermeasures by hardware relating to the reduction of the external force of a tsunami. Avoiding a tsunami by evacuation relates closely to the hard measures of the creation and improvement of evacuation routes, temporary refuge areas, and the emergency information communication system necessary for evacuation. Another possible approach to countermeasures by hardware is the creation of a "disaster-proof" town, with the aim of eliminating the danger of tsunamis in the area as a whole.

Thus, various countermeasures can be employed as hard measures in order to mitigate human and property damage. However, it is unforgettable that the countermeasures by hardware become effective on the mitigation of tsunami disasters only when used in combination of with the countermeasures by software. For example, if a tsunami larger than that assumed in hard measures attacks the protected coastal zone, the measures cannot completely prevent the tsunami from inflow. Therefore, the adequate combination with the countermeasures by software like tsunami evacuation and information is required in addition to the countermeasures by hardware. For evacuation to reduce human damage, it is necessary to secure safe refuge areas and construct hardware facilities such as evacuation route to those refuge areas. In addition, it is also necessary to learn and educate disaster prevention by using a hazard map and others which give advanced information of tsunami danger and evacuation and to prepare outdoor broadcasting equipments for the announcement of emergency information as a hard countermeasure. However, those alone are inadequate. In the evacuation at an actual emergency of tsunami attack, the persons who are evacuating should understand their own situations and take adequate actions by themselves accordingly. Furthermore, the preparation of hardware facilities is required to accomplish the smooth evacuation. Because evacuation itself relates to the countermeasure by software, nothing about it is described in this section.

From this viewpoint, it is necessary to explain the functions of the tsunami mitigation system for the regional society as a whole, considering the combination of both hardware and software countermeasures. However, it will make the explanation extremely complex. Therefore, while bearing these points in mind, the explanation in this chapter will focus on the roles and functions of hardware countermeasures.

(1) *Facilities for prevention/mitigation of tsunami invasion of inland areas*

In Japan, seven ministries and agencies involved in tsunami countermeasures prepared a “Manual on Strengthening Tsunami Countermeasures in Regional Disaster Prevention Plans” (in Japanese) in 1997 as an essential summary of countermeasures against tsunami for coastal municipalities. The manual defines the aims of tsunami disaster prevention facilities as prevention of tsunami from flowing inland and mitigation of tsunami disasters. The facilities are specifically listed on Table 3.1. This is because invasion of the inland area by the tsunami itself can be prevented and damage can be reduced by constructing these facilities. Houses, facilities, and various other fixed objects cannot be

Table 3.1 Functions and features of facilities for preventing/mitigating tsunami invasion.

Facility	Object function	Features of facility construction
(1) Tsunami seawall	Prevention of tsunami invasion of inland areas.	Construction of large-scale structure on the coastline.
(2) Tsunami breakwater	Reduction of waves attacking shore.	Construction of structures in the sea.
(3) Water gate/inland lock	Prevention of tsunami invasion of rivers and inland areas.	Construction of movable water gate at the mouths of rivers. Construction of inland locks at gaps in seawalls.
(4) River embankment	Prevention of flooding by a tsunami that runs up in a river and overflows the river embankments.	Necessary to raise the height of embankments from the river mouth.
(5) Tide control forest	Reduction of the force and run-up range of the invading tsunami.	Large forest width is necessary in order to realize the intended function.
(6) Breakwater building	Reduction of tsunami inflow behind the building, thereby reducing damage.	Construction of buildings with adequate wave resistance to withstand tsunamis along coast.

moved in the short time between a tsunami warning and a tsunami attack. In this case, preventing a tsunami invasion is the only way of reducing physical damage. Up to the present, large-scale structures such as seawalls and breakwaters have been constructed as countermeasures in order to reduce the run-up of tsunamis on land, and the effectiveness of these measures in reducing disasters has been confirmed. However, because these hardware countermeasures are extremely expensive and require much time for construction before becoming fully functional, it is not possible to apply them to every location where the risk exists. Furthermore, the frequency of tsunami attack is extremely low, and the functions of these facilities must be maintained over a long period of time. This requires continuing maintenance control, including inspections and repairs to keep them fully functional. Thus, the need to secure maintenance budgets on an ongoing basis is another point to be considered.

This chapter will present an outline of the functions and features of six types of hardware facilities: (1) Tsunami seawalls, (2) Tsunami breakwaters, (3) Water gates and inland locks, (4) River embankments, (5) Tide control forests, and (6) Breakwater buildings (Table 3.1).

(1) *Tsunami seawalls*

All of these hardware facilities are constructed along the coastline. Their function is to protect the land from invasion of sea water due to high tides, storm surges, and tsunamis. Among these facilities, the tsunami seawall (Fig. 3.8) is a facility for the primary purpose of preventing tsunami damage. This type of seawall is constructed higher than the assumed tsunami height in order to prevent overflowing by the tsunami and tsunami invasion of the inland area. This means that the "crown height" (height of the top of the seawall) increases as the assumed tsunami height increases, and the structure becomes extremely large. This causes various problems. For example, access to the shore from the hinterland is inconvenient, and the view of the ocean is blocked by the

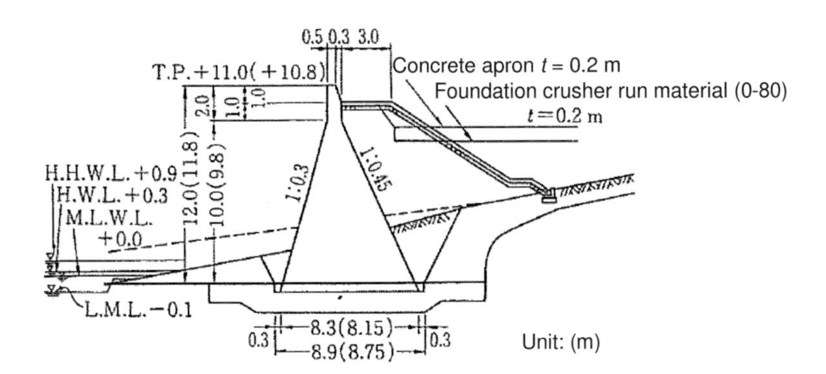

Fig. 3.8 Cross-section of tsunami seawall constructed in Okushiri Island (from Design Manual for Coastal Facilities, 2000 Ed., in Japanese).

seawall, reducing scenic beauty. Because tsunamis are extremely rare, disaster prevention facilities must be planned in combination with improvement of the everyday living environment, considering life and activities in the hinterland behind the seawall. Naturally, a tsunami may be larger than that assumed in the seawall design; in this case, the tsunami will overflow the seawall and inundate the hinterland. When planning this facility, it is necessary to consider how to drain the sea water if the tsunami overflows the seawall and inundates the hinterland. Also in the seawall structure itself, it is necessary to secure both strength against the tsunami wave force and earthquake resistance against ground motion assumed in case of a nearby tsunami.

(2) *Tsunami breakwaters (bay-mouth breakwater)*

Tsunami breakwaters (bay-mouth breakwater) (Fig. 3.9) are constructed at the mouth of a bay. Their effectiveness is to reduce the amount of water flowing into the bay during a tsunami by narrowing the entry channel into the bay, and thereby reduce the rise in the water level inside the bay. Their tsunami reduction effect increases as the area of the bay increases. Moreover, because large energy loss is created by big eddies of strong currents at the narrow entrance channel, the breakwater also has the effect of suppressing the amplification of an attacking tsunami due to resonance

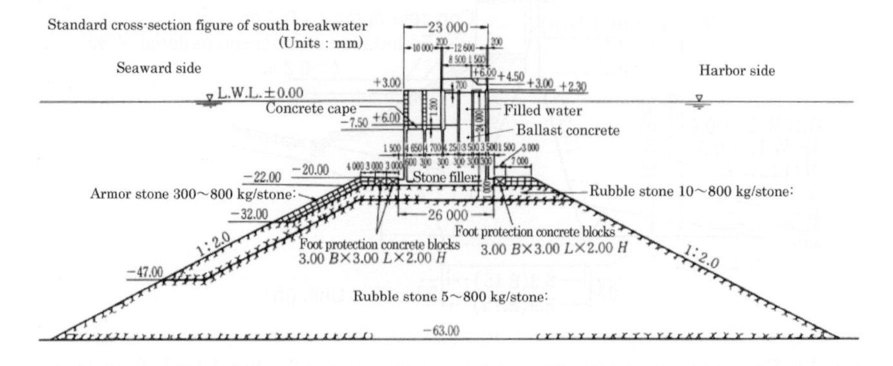

Fig. 3.9 Cross-section of tsunami breakwater at Kamaishi Bay mouth, Japan.

with the natural period of the bay. However, study of tsunami breakwaters must consider a variety of conditions, as the effect of the breakwater will differ depending on the shape of the bay, the position of the breakwater, the width of the opening in the breakwater, the period of the tsunami, and other factors. In addition, because the breakwater will also reflect the tsunami outside of the bay, the effect on the neighboring coast must also be cheked. Moreover, because tsunami breakwaters are large-scale structures which are constructed in the waters offshore, the cost of construction is extremely high, and a long period of time may be necessary for completion.

(3) *Water gates and inland locks*

A tsunami water gate (Fig. 3.10) is constructed near the mouth of a river to stop a tsunami from running up in the river. Flooding will occur if the tsunami overflows the river embankments, but this type of flooding can be prevented by stopping tsunami run-up. In this case, the water gate must be designed to adequately withstand the wave force of the tsunami. Because the water gate will reflect the tsunami, the gate must be designed to prevent an excessive rise in water level due to superposition of the reflected wave and the incoming wave.

When a vertical seawall is constructed, land locks (Fig. 3.11) are installed between the sections of the seawall to provide routine access to the coast. Under normal conditions, these are left open, but when there is a danger of tsunami attack, the locks are closed to prevent tsunami invasion of the inland area via the locks. These locks are generally closed by government employees who manage the locks and nearby residents. Therefore, it is often necessary to create a volunteer disaster prevention organization or the like for closing the locks in case of emergency, and to provide proper training. Considering the safety of persons doing the closing work, some water gates and land locks are designed to close automatically when an earthquake is sensed, or to be closed by remote operation.

Fig. 3.10 Water gate at Fukura Port in Minami-Awaji City, Japan.

Fig. 3.11 Examples of Inland locks in Japan (Left: at Shizunai Fishery Port in Numazu City, Right: at Nishina Fishery Port in Nishi-izu).

(4) *River embankments*

Tsunami which runs up against river flow sometimes flows over the embankment and causes flooding in the inland area. Such invasion of the tsunami can be avoided by constructing river embankments which are high enough to contain the tsunami. When the effect of the reflected wave from a river-mouth water gate is large, the height of embankments must also consider this factor.

(5) *Tide control forests*

Tide control forests (Fig. 3.12) have the following four functions/effects against tsunamis: (1) Reduction of intensity and energy of tsunami inflow by resistance against the tsunami; (2) Support for the "natural levee" function of sand dunes, by preventing scattering of sand and contributing to formation/maintenance of sand dunes; (3) Prevention of the spread of damage (damage to buildings, etc.) by the impact of floating objects carried inland by the tsunami ("trapping effect"); (4) Saving lives of persons who would otherwise be swept out to sea by the return flow of the tsunami, allowing them to cling to trees.

Because tide control forests cannot completely prevent the inflow of a tsunami, this measure must be studied in combination with other measures, such as city planning and the disaster prevention system. Although the quantitative effect of coastal forests in reducing the force of tsunamis is not adequately understood, various disaster prevention effects can be expected, including saving lives and trapping floating objects. For further details, refer to 2.4.

Fig. 3.12 Tide control forests.

(6) *Breakwater buildings*

Reinforced concrete buildings and other structures with adequate strength to withstand a tsunami can be arranged along the coast in order to reduce tsunami invasion of the hinterland behind such buildings. In particular, buildings which are constructed for this purpose are called breakwater buildings. Breakwater buildings can be expected to reduce the spread of damage by preventing inflow of floating objects carried by a tsunami, and also function as temporary places of refuge from the tsunami. However, study of the design standards for actual breakwater buildings is still inadequate. Items which require study include basic issues such as the necessary strength to withstand tsunamis and the optimum structural design.

(2) *Evacuation facilities*

Deaths and injury can be reduced if residents receive information on earthquakes and tsunamis and warnings/alerts and evacuate quickly. First, it is necessary to provide information communication equipment which ensures reliable communication of "tsunami information" and "evacuation information." In addition to outdoor speakers, individual receivers and similar devices that operate inside individual buildings are effective for communicating information about danger. However, evacuation information alone will not reduce deaths and injuries. In many confirmed cases, people did not evacuate even after receiving evacuation orders. In other words, transmission of evacuation orders is a necessary condition for evacuation of the local population, but it is not a sufficient condition. In combination with the information system, it is also necessary to train people to make appropriate judgments and take proper action based on evacuation information. For this, disaster education, awareness and enlightenment, and actual evacuation training are necessary. These countermeasures by software need an integrated system for the tsunami information communication under combination with software countermeasures.

Next, an environment (facilities) that enables evacuation is necessary. In other words, even assuming residents evacuate, evacuation will not be possible without safe evacuation routes and refuge areas. Without these facilities, no amount of information or training will help.

Refuge areas must be safe from a tsunami; this is the most basic point. The distance to the refuge area is equally important. In some cases, it may not be possible to complete an evacuation before a tsunami arrives. For example, in flat areas, there may be no suitable high ground nearby. In other cases, even though high ground is relatively near, the time until the tsunami attack may be too short for evacuation because the earthquake occurred close to the area. Under these conditions, it is possible to use a tall reinforced concrete building (three stories or higher) as a tsunami refuge building for temporary evacuation in an emergency. Tsunami refuge buildings must have adequate strength and height, and considering evacuees from outside, must also have outside stairs or some other means of entry. However, these temporary places of refuge are not designed for long-term refuge. Therefore, movement to some other safe refuge area after the tsunami has passed must also be considered.

Evacuation routes must enable safe passage even in emergencies. For example, when the focus of an earthquake is nearby, the earthquake ground motion that precedes the tsunami attack may topple buildings and bridges, blocking the planned evacuation routes. This condition must be avoided when designing evacuation routes. Routes that might be blocked in an emergency must be avoided from the first. If no other suitable routes exist, safety along the route must be secured or appropriate roads must be constructed in advance. Because evacuation routes are the lifeline to survival, local residents must know these routes in advance. It is also necessary to place easy-to-understand guide signs along evacuation routes for use in actual evacuations. Guide signs must be simple and easy to understand, not only for local adults, but also for children, elderly persons, and foreigners who may not know the local language.

In the towns of Kushimoto and Nachikatsuura in Wakayama Prefecture, Japan, local residents created tsunami evacuation routes with their own hands. This kind of activity by local people also contributes to the action based on one's own judgment when a tsunami strikes, as it heightens awareness of tsunami evacuation and strengthens the will to take action.

(3) *Countermeasures by planned disaster mitigation town management*

In 1997, the "Manual on Strengthening Tsunami Countermeasures in Regional Disaster Prevention Plans" was published by seven ministries and agencies involved in tsunami countermeasures at the time. Until then, tsunami countermeasures in Japan had centered on preventing tsunami disasters by constructing large-scale seawalls and breakwaters, and informing residents of the danger by issuing tsunami forecasts and encouraging evacuation. In contrast to this, the above-mentioned "Manual" aimed at mitigating tsunami disasters by comprehensively strengthening tsunami disaster prevention countermeasures using a combination of various methods. The three pillars of this program were "Disaster prevention structures," "Town-building to resist tsunamis," and the "Disaster prevention systems." This "Planned disaster prevention town management for tsunami disaster mitigation" for resistive areas against disasters requires adequate disaster prevention measures corresponding to the locality of coastal land utilization.

When tsunamis are caused by ocean-floor trench-type earthquakes, it is likely that tsunamis repeatedly attack the same area due to the same cause. Although we say "repeatedly," tsunamis are very rare, and occur at intervals of several decades or even several centuries. Therefore, it is necessary to study "town management for disaster mitigation" from a long-term viewpoint, considering foreseeable changes in the features of the regional society where a tsunami is predicted. In regions where tsunamis are expected to cause catastrophic damage, elimination of risk by drastic countermeasures for tsunami risk is most effective from the long-term viewpoint. This means preparations for disaster mitigation by infrastructure creation based on systematic urban disaster planning, including relocation to higher ground and reorganization of land use. A concrete method for this is land use management based on an evaluation of the level of risk in the region by risk assessment. Measures to be implemented considering the circumstances of the area and land use include: raising the height of the ground in the area as a whole; relocation to higher ground; and land-use regulations on high-risk areas. It is also possible to study regional plans which add the function of so-called "land dikes" against tsunamis to infrastructure such as highways and railways. For areas with a high risk of tsunami invasion, planned arrangement of groups of reinforced concrete buildings is effective. In addition to providing temporary refuge areas, these groups of buildings also reduce the range of inundation by resisting the advance of the tsunami into inland areas. It is possible to build a town that is relatively safe from disasters by using a combination of these techniques. Depending on the country, region, and local area, it is also possible to create greenbelts such as tide control forests, which are effective in mitigating tsunami disasters. For an example in the Aonae District on Okushiri Island, which suffered catastrophic damage in the tsunami following the Hokkaido Nansei-oki Earthquake in 1993, town building efforts have been carried out to avoid future tsunami damage. At the tip of the cape, where the tsunami damage was especially heavy, the residents were relocated to safer ground, and the vacated land was turned into a memorial park to convey the experience of the disaster. In the district facing the Aonae fishing port, the ground was raised to a height of 6 m to prevent tsunami invasion, creating an area where it will be more difficult for damage to occur. The arrangement of roads was also changed, considering access to the fishing port and prevention of tsunami invasion. Terraced artificial ground was constructed immediately above the landing area of the fishing port to provide a refuge area in emergencies.

In order to promote planned disaster prevention town management, agreement with the local residents and their participation are necessary. It is also necessary to consider the features of socioeconomic activity and lifestyles. Finally, if a tsunami attack is larger than expected, damage cannot be prevented completely by hardware measures alone. Therefore, it shall not be forgotten that countermeasures by hardware must be implemented in combination with those by software to minimize the damage.

3.3 Countermeasures by Software

(1) *Tsunami hazard map*

Death and injury can be avoided if residents evacuate to areas which the tsunami cannot reach. Accordingly, evacuation is the key point for countermeasures by software for tsunami disaster prevention. Because a tsunami may attack within minutes after an earthquake, quick and effective evacuation is important. To increase the number of persons who evacuate safely, it is necessary to provide appropriate tsunami disaster prevention information in advance. In particular, information on the size of the attacking tsunami and evacuation routes and refuge areas for escaping the tsunami is indispensable as tsunami disaster prevention information. This key disaster prevention information is presented in concentrated form in a drawing called a "hazard map." The tsunami hazard map is an extremely important countermeasure by software for mitigating death and injury, and is generally prepared by a governmental organization. In Japan, a manual on hazard map preparation has been prepared and issued at the national level. Local governments prepare tsunami hazard maps conforming to this manual, and distribute these maps to local residents.

The information contained in a tsunami hazard map can be broadly divided into two classes, (a) information directly related to the scale of tsunamis and the topographical features (shape) of the land, and (b) information on evacuation. In the former (a), the most important information is the range of inundation (flooding) by a tsunami attack. The time when inundation will begin, the inundation depth (water depth in flooded areas), and the flow velocity of the tsunami are effective information for establishing evacuation plans. In the latter (b), the important information is basically the evacuation routes and refuge areas.

In this connection, "dos and don'ts" and other precautions for evacuations are effective information for smooth evacuations.

When preparing a tsunami hazard map, those concerned carry out a precise numerical simulation assuming the "worst case" tsunami which is imagined in the region. However, numerical simulations of tsunamis require much work, including setting up earthquake models, preparing topographical data, and collecting information on the condition of facilities and buildings. As a result, these simulations are extremely expensive and time-consuming. Therefore, in some cases, a simplified tsunami hazard map is prepared based only on the records of past inundation. Because these simplified hazard maps only describe tsunamis that occurred in the past, it is necessary to inform users of the danger of larger tsunamis in the future.

Various people use tsunami hazard maps to mitigate damage, and the necessary information differs depending on the person. For example, in the fishing industry, users need to know if they can evacuate fishing boats to the open sea before a tsunami attacks, and if so, how far they should evacuate and how long they must wait before returning to port. This is quite different from the needs of ordinary residents. Among ordinary residents, needs of older persons are different from those of young persons. This is because older persons require more time for evacuation than young persons, and the evacuation starting time and evacuation routes may also be different. If those concerned try to prepare a hazard map that can be used by a wide range of people, the amount of information will become excessive and the map will be complicated. A hazard map with a large amount of information can be used effectively as training material for disaster prevention, but it will be difficult to use during an emergency. In other words, a hazard map that contains too much information will be "user unfriendly" in an actual emergency. Therefore, disaster prevention information must be selected considering the features of a tsunami expected to attack as well as those of the region, and composed for easy understanding during an emergency.

Reflecting the requirements and opinions of residents, interactive hazard maps are being prepared in order to provide necessary information to specific groups of residents who will use the maps. For example, interactive hazard maps for residents provide information on the best evacuation route and evacuation starting time for specific groups of residents based on their features related to evacuation (ages, genders, etc.) and information on where they live, where to evacuate, and what size of tsunami is going to attack the area. For people in the fishing industry, interactive hazard maps for fishing boats provide information on whether evacuation is possible or not, considering the size and speed of the fishing boats, conditions in the anchorage, etc.

To save lives, governments have been extremely proactive in preparing and distributing tsunami hazard maps in recent years. However, simply having a tsunami hazard map does mean it will be used. It is important that residents be fully aware of the existence of the tsunami hazard map. Maps should be posted on a wall in the home or kept where they can be taken out immediately. People who ignore the hazard map in ordinary times will not be able to use it when they need it. It is also important to give disaster prevention education and disaster training using tsunami hazard maps in local disaster prevention organizations, neighborhood associations, schools, and other groups. Regular exposure to and use of the hazard map in ordinary times will ensure that it can be used effectively in an emergency.

(2) *Tsunami disaster prevention workshop: How to use hazard maps*

Recently, decreasing evacuation rates have become a problem. These decreasing rates mean that more and more residents do not evacuate when a tsunami warning is issued. As a main cause, in many cases, a dangerous tsunami does not actually attack after a tsunami warning (or alert) is issued. After repeating this experience several times, the feeling that they themselves are the very persons who might be affected gradually decreases. As a result, people feel secure ("I'll be OK") and no longer evacuate when given disaster prevention information. If this tendency increases, escape will not be possible when a tsunami actually comes. Residents must understand and recognize that "you and your family can become victims," and at the same time, government agencies must work to ensure that residents feel that they themselves are the party concerned. For this, it is important that residents (a) understand how a tsunami can affect their own town and (b) think for themselves about how they should act in a tsunami. For (a), education is required so that residents can understand the danger of a tsunami; for (b), it is necessary to provide them with opportunities to think about the matter. What is required is disaster prevention countermeasures with residents and government acting as one, and beyond this, disaster prevention measures in which local people take the lead, and the government supports their efforts.

As a typical example of cooperative activities by residents and government, the following will introduce a "participatory-type workshop" in which a tsunami hazard map is actively used to provide disaster prevention information.

In several parts of Japan, "Tsunami Disaster Prevention Plans" are established by the workshop method. Here, a "workshop" means a method of problem-solving for obtaining rational conclusions in which a number of people offer their own ideas, as equals, in a limited time period. The fact that the participants understand the risk to the region and think about disaster prevention countermeasures in the workshop format has the following advantages:

(1) All members have a shared understanding of the risk to the region.
(2) It is possible to establish comprehensive countermeasures which reflect the opinions of many people.
(3) Because all participants feel that the countermeasures which they devise are "their own ideas," the effectiveness of the countermeasures is improved.

To understand the concrete flow of a workshop, let's look at the flow of a tsunami disaster prevention workshop which was actually held by a certain community in Japan.

(a) Know the enemy <Lecture>:

In this stage, the workshop participants learn the mechanism by which tsunamis occur, the kind and scale of damage that tsunamis cause, methods of tsunami disaster prevention, and past tsunami damage in their own region.

(b) Know the region <Workshop> (Fig. 3.13):

(1) Participants plot their own houses on the map.
(2) Participants plot community assets on the map (these include Jizo (roadside stone statues of Buddhist saint), shrines, assembly areas, parks, pump stations, areas where people gather, stores, wells, important properties of the town).

(c) Know the damage <Workshop and lecture>

<Workshop>

(1) The estimated damage shown on the tsunami hazard map is transferred to the map mentioned above.
(2) Participants understand the damage to their own homes and the assets of the community.

<Lecture>

(3) Participants learn actual kinds of damage that may occur, the time until arrival of a tsunami, and methods of preventing damage.

Fig. 3.13 To know local features.

(d) Thinking about countermeasures <Workshop>:

Participants think about the kind of countermeasures that should be taken for the estimated damage in the region.

(1) Participants write down necessary countermeasures on Post-It cards (one countermeasure on one card).
(2) Next, they present their own ideas, and place their Post-Its on a large sheet of paper. Each participant presents one idea in each round of the presentation. This process is continued one by one until all of the Post-Its are used. However, when participants think of another good idea during the presentation, they can write it down on a new Post-It.
(3) All the ideas thus collected are then structuralized.

(e) All participants share the results of the workshop (Fig. 3.14).

<Items to prepare for a workshop>

(1) Map of the region (can be prepared easily and used by printing a Google Map on large paper)
(2) Ordinary writing pens (one for each participant)

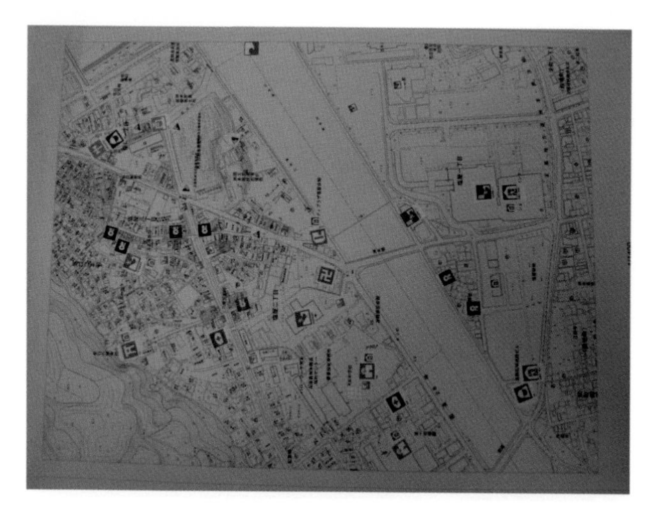

Fig. 3.14 Local hazard map prepared by a workshop.

(3) Thick tip pens (four colors)
(4) Post-Its (three colors)
(5) Office supplies (glue, cellophane tape, etc.)
(6) Hazard map

One obstacle to establishing disaster prevention plans in Asian countries using this workshop format is that these countries (or regions in the countries) do not necessarily have tsunami hazard maps. However, if you believe that you cannot create a disaster prevention plan because you do not have a hazard map, you may wait for a long period of time without acting. Without information from the government or experts regarding the possible size of a tsunami on the coast, it is not possible to prepare hazard maps for the region. However, for example, it is possible for local people to gather "height-above-sea-level" information for their area. Needless to say, experts or the government must provide the tsunami hazard information, but more importance shall be placed on the autonomous efforts by local residents who will be affected by an actual tsunami. These efforts might include, as mentioned above, measuring the height above sea level at various locations in order to know the safest places for evacuation and the danger spots in the area (Fig. 3.15).

Fig. 3.15 Sign showing the height above sea level.

(3) *Disaster prevention pictograms: Informing strangers to the region about local risk and refuge areas*

When considering disaster prevention countermeasures, items which must be protected include three elements: life, property, and living condition. Naturally, priority is in the order of "life > property > living condition." In tsunami disaster prevention, it is difficult to protect all of these through employing countermeasures by software alone. Thus, in order to consider comprehensive countermeasures for a region, it is necessary to implement a skillful combination of countermeasures by software and those by hardware. Regardless of the level one thinks about in the priority order of "life > property > living condition," the first step in tsunami disaster prevention countermeasures is to know the risk to the region.

Publication of hazard maps is a general method of informing people of the risk in the region. However, it is difficult to inform all those who may be affected by a tsunami using only hazard maps. Examples include tourists and others who do not live in the region and people who have no interest in disaster prevention. An "evacuation guidance system" using disaster prevention pictograms is an effective means of informing people who have little knowledge of the region, centering on tourists, about the tsunami risk in the region and the minimum response one must make when attacked by a tsunami to save one's own life.

"Pictograms" are sign symbols using picture writing. A typical example is the picture of a running man on the sign for "emergency exit" at building evacuation doors. These pictogram signs communicate their messages using a combination of color (blue: mandatory action or instructions, yellow: caution or danger, green: safety or evacuation notice) and shape (circle: forbidden or mandatory action, triangle: warning, square: information). As one advantage of pictograms, it is possible to communicate the meaning to children who cannot read and foreigners who do not understand the local language. Among the pictograms used for tsunami disaster prevention, various countries currently use different kinds of graphical symbols (Fig. 3.16). However, the Great Indian Ocean Tsunami of 2004 spurred moves to standardize these symbols. As shown in Fig. 3.17, three Japanese designs ("tsunami alert," "tsunami refuge area," and "tsunami refuge building") have been proposed to the ISO.

Tsunami pictogram under study by UNESCO

Tsunami pictogram used on the West Coast of the United States

Fig. 3.16 Examples of various tsunami pictograms.

■ Tsunami alert

■ Tsunami refuge area

■ Tsunami refuge building

The color of these graphical symbols is specified in JIS Z 9101 ("Safety colors and safety signs-design principles for safety signs in workplaces and public areas").
Please refer to the following Munsell values:
Safety color Green: 10G 4/10, yellow: 2.5Y 8/14
Color contrast Black: N1, white: N9.5.

Fig. 3.17 Tsunami pictograms proposed by Japan (Disaster Prevention Pictogram Research Council, 2005).

An evacuation guidance system using disaster prevention pictograms comprises two parts: an "education system" regarding the risk in the area and evacuation actions, which is used before a disaster; and an "emergency information system" which provides information when a tsunami actually occurs. The "education system" consists of the following elements:

(1) A system of tsunami risk signs, which shows the danger of waves and the history of tsunami damage in the region;
(2) A system of evacuation route/refuge area signs, which shows refuge areas and tsunami refuge buildings for use at a tsunami occurrence as well as evacuation reports and the safe height.

Efforts to develop an evacuation guidance system using disaster prevention pictograms in a town as a whole are important for creating an area that can survive a tsunami disaster. This may include "education-type signs such as hazard maps" in places where many people gather, "refuge area signs" on safe high ground, and "signs showing the direction to refuge areas" at street junctions. An example of the development of a pictogram system in a town is shown in Fig. 3.18.

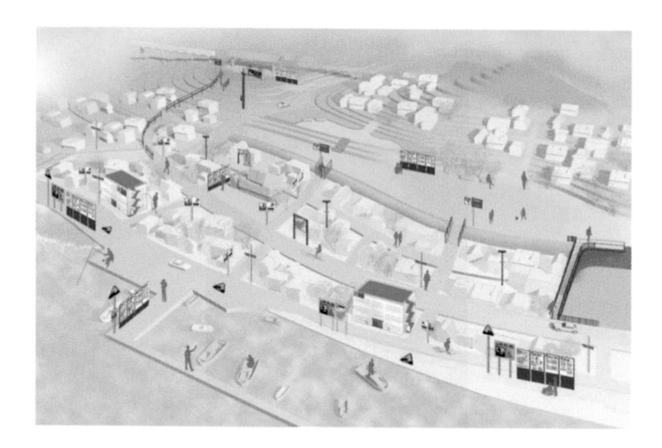

Fig. 3.18 Tsunami disaster prevention pictograms deployed in a town (Disaster Prevention Pictogram Research Council, 2005).

(4) *Construction of stone monuments*

Learning about past disasters is important for preventing and mitigating disasters of all types, and not only tsunamis. We must know about the disasters which have occurred in the past and the damage such disasters have caused and learn from them. For example, if a tsunami has occurred in an area, the actual experience is important information that should be left to our descendents. (For example, residents may have experienced a tsunami attack immediately after feeling an earthquake, and this was followed by repeated attacks by the second and third waves.) The wisdom and traditions born out of this kind of disaster experience are called "disaster culture." Handing down this culture from generation to generation is indispensable for mitigating the damage in future disasters.

However, in comparison with earthquakes and typhoons, the frequency of tsunamis is low. For this reason, the memory of the disaster fades, and formation of a disaster culture is difficult. Therefore, efforts to pass on the experience of past disasters and ensure that this experience is useful for disaster prevention and mitigation are necessary. One such effort is the construction of stone monuments.

Many stone monuments for past tsunami disasters have already been constructed (Fig. 3.19). The damage caused by the tsunami and the

Fig. 3.19 Ryoishi Tsunami Monument in Kamaishi City, Iwate Prefecture, Japan. This monument describes the tragedy of the Showa Sanriku Earthquake Tsunami of 1933.

lessons learned are carved in the stone and preserved as a monument in order to pass on the disaster culture. These stones convey the tragedy of the tsunami disaster, but at the same time, also they provide precious information on the damage at the time. It is possible to investigate past disasters from the distribution and content of these stone monuments. Because there is little documentation on many of the tsunamis that occurred in ancient times, the existence of these monuments has made an important contribution to research on historical tsunamis.

Depending on the region, there are cases where stone monuments for tsunamis have been buried in ground and are no longer visible to the area's residents. There are also many cases where residents do not know the content written on the stones. The people of the past made deliberate efforts to create a disaster culture. To ensure that this disaster culture is not lost, it is necessary to protect local stone monuments and make efforts to communicate their content. This is important not only in the sense of preserving historical assets, but also as a key element in the countermeasure by software for improving the disaster prevention capabilities of the region.

References

Disaster Prevention Pictogram Research Council, Tsunami Disaster Prevention Pictograms 2005, Disaster Prevention Pictogram Research Council, 2005.

Uhana, M. (2002): Tsunami stone monuments and memorial stones (including tumbstones) in Sanriku coastal areas – Misawa-Shi in Aomori Pref. to Iwaizumi-Cho in Iwate pref.-, Research Report of tsunami Engineering, Vol. 19, Part 2, pp. 1-73 (in Japanese).

3.4 Transmission of Experience and Education

(1) *A lesson called "tsunami tendenko"*

Perhaps the most impressive lesson from tsunami disasters in Japan is the one called "*tsunami tendenko*." It seems that this lesson was widely understood after the Meiji Sanriku Earthquake Tsunami of 1896, which resulted in more than 22,000 dead and missing victims in an area of northeastern Japan on the Pacific coast. The meaning of the expression "*tsunami tendenko*" is "In tsunami, everybody is on one's own": that is, everybody, including parents and children, should run for their own lives separately, without relying on other persons. Although this sounds cold-hearted, it includes the hope that families can avoid dying together, resulting in the end of the family line. In turn, the idea that each individual should protect his or her own life means the number of survivors in a village or region will be increased in that way.

In other countries, similar examples were also reported in the 2004 Great Indian Ocean Tsunami. The testimony already described in this book (1.2(5)) is truly an example of "*tsunami tendenko*." A certain Mr. Silver was on a train with his parents and three children when the tsunami struck. He put only one of his children on his back, and the two miraculously survived. As he climbed out of the train window, he pushed away the people clinging to his legs. He later said, "The people clinging to my legs might have been my own parents, but I was determined to save at least this one child."

On the other hand, in the Hokkaido Nansei-oki Earthquake Tsunami in Japan in 1993, few people thought that the tsunami would attack so quickly. Many of those who died waited to escape with their entire families, delayed escaping in order to load family belongings into their cars, returned to their homes to collect valuables, or stopped to warn others so they could escape together. Here, the lesson of "*tsunami tendenko*" was not applied. Evacuation from a tsunami is that difficult and unforgiving.

(2) *Significance of handing down the story of tsunami disasters from generation to generation*

Many of the tsunami experiences of individuals which are handed down, and the lessons based on those experiences, are unique cases. There is also a high possibility that this knowledge is not scientific. Nevertheless, handing down the stories of tsunami experience from generation to generation is extremely significant because actual objects, photographs and videos, memoirs, and stories of experience related to disasters have enormous power to move people's hearts and eloquently communicate the flesh-and-blood feelings of the victims in those events. When we hear these stories of disasters, we are moved through the experience of others and can feel the helpless fear and sorrow of persons caught in a disaster. This is difficult with replicas and fiction. Monuments for the repose of victims and memorials of the disaster are also extremely important, but the message that they communicate is somewhat different from the stories of actual experiences that were directly handed down. This is an important reason why handing down stories of disasters is so significant.

The authors asked high school students living in tsunami-prone areas about the transmission of tsunami knowledge from older generations. As a result, it was found that 78% of the students who had been living in the area since birth had heard traditional stories about earthquakes and tsunamis; on the other hand, only 21% of students who had moved in from other areas had heard these stories. In an extremely large number of cases, the students heard these stories from a parent (77%) or grandparent (56%), followed by elementary school teacher (42%) and junior high school teacher (45%). In other words, many children who are raised in tsunami-prone areas receive some kind of tsunami knowledge through stories in families that have actually experienced a tsunami disaster. However, in the case of children who moved in from other areas, nobody in the family has experienced a tsunami, and there are few opportunities for transmission of knowledge by family members. From a different viewpoint, this kind of oral transmission occurs either in the family, or at most, in a narrow limited range of the local society.

Furthermore, in Japan, the number of children who are born and grow to adulthood in one local area is steadily decreasing. Children frequently move away from their native area and relocate in new places because their parents change workplaces, or the children themselves go on to universities, find work, or change workplaces. Children from the mountains may move to coastal areas and vice versa. How should children learn about the various disaster risks in a region, including tsunamis? And after they have their own families, how should they teach their children? All children must receive an equal education in these matters, and opportunity for this education must be guaranteed continuously, even when children move to other areas.

For this, it is fundamentally important that persons who have experienced disasters in the past hand down their own experience and lessons to many people in new generations. The zeal of people in disaster areas to "pass on their terrible experience to future generations" is the most important driving force for a continuing oral tradition of disaster stories. At the same time, when victims describe their own painful experiences to other persons, this oral transmission has a healing effect for the victims and gives them the energy to make a new start, while also rebuilding their relationship with society. However, over long periods of time, it is very difficult for survivors to maintain the same strong emotions which they experienced during an actual disaster. At the individual level, feelings easily fade. The effect of individual efforts is also very limited. Accordingly, it is important to preserve the feelings and impressions of people regarding the affected area at the time of the disaster, not only through the efforts of individual disaster survivors, but also at the level of society as a whole. It is essential for the survivors to reconfirm their impressions during the disaster from time to time, and to communicate these impressions to persons who did not experience the disaster. It is also important to increase the effectiveness of the oral transfer of disaster experience, and to emphasize its importance, by organizing persons and groups that are separately trying to pass on disaster experience in various parts of the world toward the common objective, and speak to the world as a whole. The same can be said of all disasters, not limited to tsunami disasters. Based on this recognition, the International Disaster Transfer Live Lessons Network (TeLL-Net) was established in Kobe in January 2006.

Activities to transfer disaster experience do not necessarily require large financial resources. It is hoped that the developing countries, many of which have suffered heavy damage as a result of natural disasters, will use various kinds of ingenuity to ensure effective transfer of disaster lessons.

(3) *Exhibition facilities as media to transfer disaster lessons*

Exhibition facilities are a method which makes it possible to communicate the memories and records of disasters permanently on an everyday basis, and have an effect on a wide range of society. Exhibition facilities have various kinds of functions. For example, some are strongly colored by repose for the disaster victims, while others provide a pleasant learning experience about natural phenomena and the mechanism of disasters, and some convey the threat of disaster by exhibiting primary materials and objects in their original form, and still others also have unrelated functions such as introducing local history or providing tourist information. Exhibition facilities are full of variety corresponding to types of disasters. The following presents individual examples, limited particularly to tsunami and earthquake disasters.

[Okushiri Island Tsunami Museum] (Japan)

In July 1993, Okushiri Island was hit by an earthquake, which was followed immediately by a tsunami and fire. A total of 198 persons died in this tragedy as described in 1.3. In order to keep this disaster alive in memory and transmit the lessons to future generations, and to show appreciation for the support which the island's recovery efforts received from throughout Japan, the Okushiri Island Tsunami Museum was opened in Aonae District. Aonae suffered the most severe damage due to the tsunami and fire, and the houses in the low-lying Aonae District were not rebuilt. However, in addition to the museum, a memorial greenbelt park and stone monument were also constructed.

This museum includes a large space for a monument for the repose and memory of the victims. In addition, three-dimensional models (48 in total) recreate Okushiri Town from its birth to the disaster and recovery, and a documentary film shows from the mechanism of the Hokkaido Nansei-oki Earthquake to the disaster and recovery process. Exhibitions also highlight the attractiveness of the island, including its history and natural beauty.

On the other hand, there are few objects which give a feeling of the danger during the disaster or the emotions of the victims. In recent years, photographs showing conditions immediately after the tsunami attack and poems and essays about the horror of the tsunami by elementary and middle school students have been added to the exhibits. For more information, refer to the Okushiri Island Tourist Association website: http://www.unimaru.com/fukkou_tsunamikan.html (as of May 8, 2007).

[Great Hanshin-Awaji Earthquake Memorial Disaster Reduction and Human Renovation Institution] (DRI; Japan)

The Great Hanshin-Awaji Earthquake in January 1995 claimed 6,434 lives. The earthquake, centered near Kobe, was a direct hit on a densely populated modern city. Many lessons were learned as a result of this disaster, including damage to the urban infrastructure which greatly exceeded predictions, the limits of governmental response, and the problems of a local society with an aging population. As a result, Japan has shifted its priorities from "conquering nature" to "reducing damage while coexisting with nature."

Following the disaster in 1995, the "Great Hanshin-Awaji Earthquake Memorial Disaster Reduction and Human Renovation Institution" (DRI) was opened in order to transmit the experience of the earthquake disaster to future generations and ensure that its lessons can be used in the future. The activities of the DRI include housing primary materials recovered from the disaster site, recreating the earthquake through video and audio technology, and displaying articles and video images that show the conditions from the occurrence of the disaster until reconstruction. A distinctive feature of the DRI is operation with a strong consciousness of the ties between residents of the disaster site and governmental organizations.

In particular, primary materials were collected systematically by ordinary citizens and others from an early stage in the recovery and reconstruction process. The DRI houses approximately 160,000 such items, which are available for inspection. Exhibits include 800 articles, photographs, and videos, such as videos of the urgent emergency lifesaving efforts, photos of totally-destroyed houses, and memoirs of survivors. This museum also provides oral descriptions of the actual experience and lessons of the earthquake by volunteers in a Recital Corner, where earthquake survivors talk about their experiences.

Activities are not limited to oral transmission of actual disaster experience. Because the DRI has a strong awareness of the need to correctly generalize these individual experiences as knowledge that can be shared by the society as a whole, its activities include training of government employees throughout Japan, with the aim of providing disaster prevention knowledge in a systematic and comprehensive manner, and practical disaster prevention research by young researchers.

In addition to the DRI, various activities are carried out widely throughout the Hanshin-Awaji region in order to pass on the experience and lessons of the earthquake disaster, not limited to either the public or the private sector. These include monuments, preservation of ground faults, events, and other regional disaster support activities. For more information, see the DRI website: http://www.dri.ne.jp/ (as of May 8, 2007).

[Adapazari Earthquake Culture Museum] (Turkey)

In August 1999, an earthquake with its focus in northwestern Turkey caused massive damage, including more than 17,000 deaths. The city of Adapazari, which is located near the center of the earthquake, was also damaged. Following this tragedy, the Earthquake Culture Museum was opened in the city to convey the experience and lessons of the earthquake disaster to future generations.

The exhibits in this museum include photographs taken immediately after the earthquake, a recreation of an ordinary family's living room after the quake, before-and-after photographs of main points in the city, records of past earthquakes (the same region has suffered similar earthquakes at intervals of about 20 years), and newspapers from the time of the disaster. Persons living in Adapazari at the time also describe their sensations during the earthquake to museum visitors. Outstanding entries in children's essay and drawing contests are exhibited, and disaster prevention education is given for children.

[Pacific Tsunami Museum] (Hawaii)

Hawaii has been attacked by tsunamis several times in the past at intervals of approximately 60 years, and has suffered serious damage. In particular, 159 persons died in a tsunami in April 1946, and 61 died in a tsunami in May 1960. In 1993, a certain tsunami survivor suggested collecting tsunami stories and photographs, and persons who agreed with this idea began this in voluntary activities. This group acquired a permanent space in 1997, when a bank donated a building that it had been using as a branch. The building was then renovated, resulting in the museum in its present form.

The museum contains a compact exhibition on past tsunamis and the mechanism of tsunamis and shows videos of interviews with tsunami survivors. The museum also houses, organizes, and preserves various types of documents, including scientific papers, newspaper articles, and others, relics such as a bent parking meter, photographs and videos, and other primary and secondary materials, and makes these items available to researchers. In particular, it is active in collecting interview records and memoirs of tsunami survivors and tsunami witnesses. Because persons who have direct experience of past tsunamis are growing older and passing away each year, the museum also actively holds annual gatherings of those concerned, as well as other events.

The administrative system is widely supported by local people. Virtually all of the administrative staff and exhibition guides are volunteers, such as university professors who have retired from active careers and others. Many museum visitors are local residents, including school children doing outside study. Unlike museums under government control, the scale and organization of the Pacific Tsunami Museum are small, but it is an extremely interesting example because it actively conducts programs that give priority to its relationship with the community. Part of the contents of the Pacific Tsunami Museum can be viewed at the museum's website: http://www.tsunami.org/.

Museums related to the Great Indian Ocean Tsunami

Various areas that suffered damage in the Great Indian Ocean Tsunami in December 2004 plan to construct exhibition facilities in order to pass on knowledge of the disaster. For example, in Sri Lanka, construction of facilities in the disaster area is being studied under the leadership of the National Museum. These facilities would include a memorial to the tsunami victims, educational activities related to tsunamis, preservation of damaged items, etc.

(4) *Uniqueness of traditional descriptions and the diversity of tsunamis*

Many residents of areas that have frequently suffered tsunami attacks in the past have knowledge about tsunamis, for example, that the tide goes out before a tsunami comes, a rumbling "sea noise" can be heard when a tsunami approaches, and the first wave of a tsunami is small, but later waves become progressively larger. Investigation has shown that virtually all of this knowledge is based on the person's own experience of a tsunami, or on stories of tsunami experience passed down from older generations. Being based on actual experience, this kind of knowledge is not mistaken. However, readers should fully understand that this knowledge only describes one local aspect (uniqueness) of a particular tsunami disaster.

For example, it is generally said that "the tide suddenly goes out before a tsunami hits, and you can see parts of the sea bottom you've never seen before," and one should flee if this happens. In the Great Indian Ocean Tsunami, Simeulue Island in Indonesia was extremely close to the center of the earthquake, but the number of deaths was quite small. At Simeulue Island, in addition to the fact that high ground exists near the shore, the events during a tsunami in 1907 have been passed down to the present day in a folk song. The residents immediately evacuated to high ground when they saw the tide go out, and as a result, they survived. However, this traditional belief is not always correct. As explained repeatedly in these pages, a tsunami attack may sometimes begin with a rising "run-up" wave. The traditional wisdom says "if the tide goes out, run." However, if this is expanded to mean "if the tide doesn't go out, don't run," and a tsunami begins with the run-up wave, the result can be death. Compare the two expressions, "before a tsunami comes, the tide will go out" and "after the tide goes out, a tsunami will come." There is just a slight difference in these two expressions, but only the second is correct. The first is dead wrong.

To give one more example, residents who experienced the Chilean Earthquake Tsunami of 1960 stated that "the tsunami became larger in the inner bay at Kesennuma Bay, but was not large in the middle part of the bay because it was where the tsunami just passed through." Certainly, film from the time confirms that the flow was extremely calm in the middle part of the bay. Thus, the residents' statements were correct. However, it is difficult to say that this is a knowledge generally applicable to every tsunami. The amplification characteristics of a tsunami in a bay are determined by the relationship between the spatial scale (width, mouth-to-back distance) of the bay and the period (wavelength) of the approaching tsunami. Accordingly, even in the same bay, the amplification of the tsunami can occur in the middle part of the bay rather than at the back, depending on the approaching tsunami. In fact, in the Hokkaido Toho-oki Earthquake Tsunami of 1994, the height of the tsunami increased from the mouth to the middle part of the bay, and then decreased at the back of the bay, even though the scale of the bay was similar to that at Kesennuma Bay.

Tsunamis have individual uniqueness. Every tsunami is different from others, and even the same tsunami will present a different face at different places. If knowledge based on actual tsunami experience is passed on, the persons who receive that knowledge may unfortunately believe that it is a generally applicable knowledge. If simply trusted, past examples and traditional beliefs can have the opposite effect when quick decisions are necessary, and can cost people their lives. In order to avoid this kind of mistaken judgment, it is important not simply to accept and learn past examples and traditional beliefs. In combination with this, an understanding of the general physical characteristics of tsunamis is also necessary.

(5) *Current status of tsunami disaster prevention education and its effects*

In tsunami-prone regions in Japan, evacuation drills assuming tsunamis are conducted periodically. In recent years, disaster prevention education has been included in school education, irrespective of the conventional course framework. The Disaster Prevention Education Challenge Plan is a program recently initiated in Japan to introduce model cases of advanced local efforts and assist the implementation of disaster prevention education at the local level. Through various means such as this plan, diverse examples of tsunami disaster prevention education efforts have been reported. These examples can be classified in the following three categories, depending on the mode of implementation and purpose. Among the diverse examples of disaster prevention education, some have specialized purposes, focusing on one of the following three elements while others are comprehensive efforts that include some or all of these elements.

Lecture type: In this type, study is based on talks about experience by persons who have actually experienced tsunamis and lectures by experts. The content emphasizes the history of tsunami disasters, the mechanism by which tsunamis occur, the nature of tsunamis, and the damage and fear that tsunamis cause.

Workshop type: This is a system in which the participants themselves discover the weaknesses of their area in case of a disaster by doing map exercises, preparing hazard maps, and participating in various other workshop activities.

Practical training type: The purposes of this type of training are to ensure the preparedness in everyday life and to heighten the awareness of the importance of countermeasures by simulated experience of tsunami disasters. This includes evacuation drills, emergency distribution of hot meals, etc.

In order to investigate the effects of these types of tsunami education, attitude surveys of the participants were conducted before and after education. The results were as follows.

[Lecture type]

The subjects of this study were a total of 309 1st and 2nd year high school students in a tsunami-prone area. A one-hour tsunami disaster prevention course was held, focusing on the mechanism of tsunami occurrence and propagation, tsunami damage, tsunami countermeasures, and key points for surviving a tsunami. The effects of the course were investigated, showing changes that included a greater understanding of the history of tsunami disasters and the seriousness of the damage in the town where the students lived, and a strong awareness that the town might suffer enormous damage from a tsunami in the near future. In other words, the tsunami disaster prevention course raised their awareness that they themselves are the very persons who might be affected." Unfortunately, the study did not investigate the extent to which this increased awareness was maintained.

[Workshop type]

In holding a workshop to prepare a tsunami disaster prevention map, the participants confirmed certain effects, including understanding of the vulnerability of their area to tsunamis, recognition of the necessity of evacuation, sharing of information, understanding of tsunami damage, etc. After the workshop, an earthquake actually occurred. Therefore, a follow-up study on the workshop participants was conducted. As a result, increased preparedness and awareness of tsunami disasters could be seen, for example, in checking evacuation routes and refuge areas, discussions with neighbors, and checking dangerous areas. Thus, it appears that actual preparedness is further improved if residents who have received disaster prevention education actually experience a disaster themselves.

[Practical training type]

A disaster prevention camp for elementary school students was held, and various training activities were carried out, including a classroom evacuation drill, firefighting drill, first-aid class, emergency distribution of hot meals, etc. The results of a follow-up study on the effects showed an increase in opportunities for discussion of disasters in families.

From the examples presented above, certain effects of implementing tsunami disaster prevention education can be confirmed. For example, people's awareness is increased, and the content of the education is reflected in the preparedness in everyday life. However, tsunami disaster prevention education tends to be limited to increasing the awareness of only the participants, and this does not extend to the area as a whole. Regarding event-type disaster prevention education, many people have the opinion that the preparations for events requires too much time and work, and it is difficult to secure the time and money to hold regular events. An important feature of tsunami disasters is their low frequency. When a valuable educational opportunity ends as a one-time event, participants' once-heightened awareness gradually fades. Therefore, continuing implementation is desirable.

In Japan, tsunami disaster prevention education is currently given mainly to people who live in areas with a risk of tsunami attack. This does not mean that it is not necessary to educate people who live in inland areas far from the sea, where a tsunami attack is physically impossible. To date, efforts have simply short-handed. Wider education is necessary because people who live far from the sea also visit the coast on trips and for recreation, and there is a possibility that they may encounter a tsunami. In fact, in the 2004 Great Indian Ocean Tsunami, people from landlocked countries in Europe and a large number of people from countries that do not suffer tsunami attacks died in the beach resort areas of Thailand. Even though Britain has no experience of tsunami disasters, children are taught about tsunamis in elementary school geography classes. One 10-year old British girl who remembered this lesson saved the lives of several hundred tourists at Phuket in Thailand during the Great Indian Ocean Tsunami. This example clearly shows the importance of educating a wider range of people.

(6) *Danger of over-reliance on disaster prevention information*

Needless to say, the most important measure for surviving a tsunami disaster is "quick evacuation," including preparedness for such evacuations. An extremely important part of preparedness is accurate information and advanced information transmission. Unexpectedly, the advanced information society of recent years tends to have the opposite effect and hinder quick evacuation. For example, on May 26, 2003, a magnitude 7.0 earthquake occurred, with its focus off the coast of Miyagi Prefecture, Japan. Along the Sanriku coast in Iwate and Miyagi prefectures, a seismic intensity of 4 to slightly less than 6 was observed, but 12 minutes after the earthquake, the Meteorological Agency announced that there was no danger of a tsunami. (Because the center of the earthquake was deep, a tsunami did not occur.) However, this experience revealed one problem: Only a mere 1.7% of residents evacuated during the 12-minute blank period before the announcement was made informing there was no danger of a tsunami. In other words, almost no residents of the area began evacuations. This was not necessarily due to a lack of awareness about tsunamis. Rather, almost all residents had a high awareness of tsunamis and thought that a tsunami might attack immediately after the earthquake. However, after the tremors stopped, people waited for tsunami information and evacuation information, delaying their evacuation until this information was announced. Many other similar examples have been reported showing a negative effect of the advanced information society.

In summarizing this section, the author wishes to emphasize that, in order to save yourself from a tsunami, you must not depend blindly on tsunami information. A thoroughgoing “awareness that you yourself must protect your own life” and an understanding of the importance of “quick evacuation” are essential. The awareness that you must protect your own life is not developed by disaster information, but rather, by the traditional stories handed down from former generations and lessons of people who have experienced disasters. In addition, we must not forget the need for correct knowledge about disasters.

References

Abe, Y., Imamura, F., and Ushiyama, M.: Preparation of a tsunami disaster prevention map by resident participation and related issues, Abstracts of Lectures, Technical Research Presentation Conference of the Tohoku Branch of the Japan Society of Civil Engineers, FY2003, pp. 170-171, 2004. (in Japanese)

Disaster Reduction Alliance 2005, Introduction to the Adapazari Earthquake Culture Museum, January 18, 2005, Kobe, Japan. (in Japanese)

Executive Committee of the Disaster Prevention Education Challenge Plan: http://www.bosai-study.net/ (in Japanese)

Karatani, Y., Koshimura, S., and Shuto, N. (2003): Research on systemization of disaster prevention knowledge for sustainable disaster prevention education in tsunami-prone regions, Proceedings of Coastal Engineering, Japan Society of Civil Engineers, Vol. 50, pp. 1331-1335. (in Japanese)

Katada, T., Kodama, M., Kuwasawa, N., and Koshimura, S.: Current status and issues of tsunami disaster prevention examining the evacuation actions of residents – Based on a survey of the 2003 Miyagi-oki Earthquake/citizens' awareness in Kesennuma City – Journals of the Japan Society of Civil Engineers, No. 789/II-71, pp. 93-104, 2005. (in Japanese)

Senarath Wickramasinghe: Proposed Disaster Museum, Materials of the General Meeting for Establishment of the International Disaster Transfer Live Lessons Network (TeLL-Net), January 20, 2006, Kobe, Japan.

Yamashita, F. (2005): The Terror of Tsunamis, Tohoku University Press, p. 249. (in Japanese)

Yoshida, K., Sugawara, M. et al.: Efforts at a disaster prevention camp in Tsukidate Elementary School, Kesennuma City, Collected Abstracts of Lectures at 24th Scientific Conference of the Japan Society for Natural Disaster Science, pp. 147-148, 2005. (in Japanese)

For example, http://en.wikipedia.org/wiki/Tilly_Smith.

3.5 Regional Efforts to Implement Comprehensive Tsunami Countermeasures

In preparation for earthquakes and tsunamis, it is important that the national government, local governments, the private sector, and the residents of the region respectively implement comprehensive disaster mitigation countermeasures autonomously and continuously, on an everyday basis, in order to heighten the disaster-prevention capacity of the region, based on the assumption that "a wide-area disaster is certain to occur." Furthermore, these countermeasures must integrate both hardware and software measures.

Readers should notice that we do not say "disaster prevention countermeasures"; rather, we say "disaster mitigation countermeasures."

There is always a danger that an earthquake/tsunami will exceed predictions, and it is difficult to implement countermeasures that will reduce damage to zero. Therefore, we use the term "disaster mitigation" in order to clearly indicate that the purpose of countermeasures is to reduce damage.

Efforts will be made to reduce the human damage by a tsunami to as close to zero as possible. The key point here is to create a system which ensures that evacuations are (or can be) carried out effectively. A precondition for this is that all persons who have some relationship with the region have a strong awareness, firstly, of their responsibility for their own lives, and then, of assisting each other. ("All persons" naturally mean the actual residents of the region, but also persons who work in the region and tourists and others who are only in the region temporarily, including foreigners.) This is a spirit of "self help" and "mutual help." To foster a spirit of "self help" and "mutual help," the national government and local governments must work to increase social acceptance and alertness, while steadily continuing appropriate communication of risk with residents. This is done by:

Making it clear to the area's residents what can and cannot be done with public support;
Providing risk information to residents, etc.

Local governments must study various matters in detail and implement an appropriate response based on the reality that, as in the past, some residents will not evacuate after an earthquake even if a tsunami warning is issued. Therefore, matters for study by local governments include:

Promotion of active action by residents;
Direct provision of information, for example, using cell phones;
Communication of information, including tsunami forecasts, not only to residents, but also to tourists and foreigners (who do not know the area and may not understand the language)

To use a word corresponding to the terms "self help" and "mutual help," which express the spirit of the area's people, the spirit of the national government and local governments can be expressed by "public help."

This section will introduce cooperative efforts of the related governmental organizations in the region and tsunami disaster mitigation countermeasure plans established autonomously by local governments.

(1) *Vulnerability of the Kinki coastal region to tsunamis*

Japan's Kinki region has population of 14 million persons and total annual gross regional production of US$650 billion, which the region is comparable in scale to the country of Canada. In this region, major cities are built on soft ground with poor resistance to earthquakes, and the central part of the region is a large "zero-meter" zone (zero meters above sea level), where inundation damage can easily spread. Moreover, the region's coastal industrial zones have a high concentration of facilities that handle hazardous materials. Considering these various factors, a large-scale disaster is a serious concern.

Because the ocean bottom in the south of the Kinki region has been the site of trench-type Tonankai and Nankai earthquakes at intervals of 100 to 150 years, a gigantic earthquake of M8.6 class is considered to be almost certain by the middle of the 21st century. After such an earthquake, a tsunami might reach the region within several minutes at the earliest, or in about two hours at the latest. Based on these assumptions, enormous human and physical damage resulting from a large earthquake and large tsunami are feared in the coastal areas of the Kinki region, where the population is concentrated and property has accumulated close to the focus of an assumed earthquake. In particular, attack by a tsunami exceeding 8 m is predicted along the southern coast (Fig. 3.20). Because this area will be near the focus of the earthquake, the time until the tsunami arrives will be short. The human damage by a tsunami in this area will be huge. In the worst case, the estimated death toll is 1,400 to 3,300 persons.

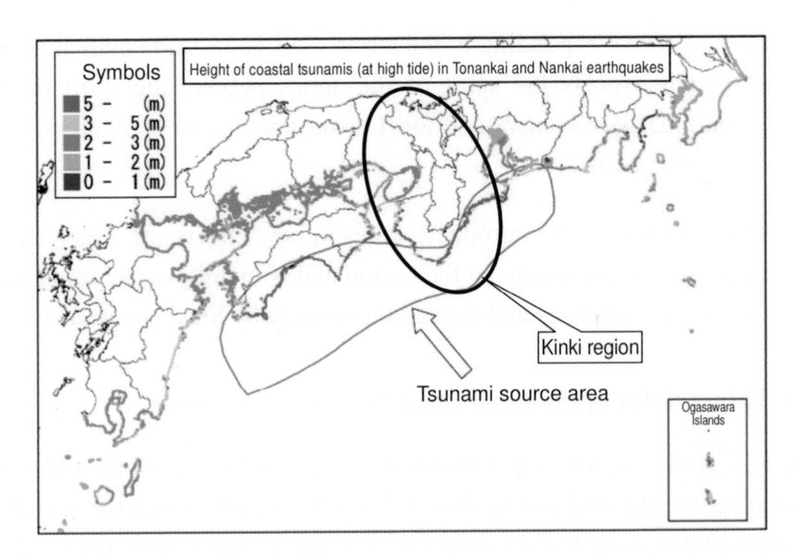

Fig. 3.20 Height of tsunamis caused by Tonankai and Nankai earthquakes (source: Central Disaster Management Council).

In the Kinki coastal region, the back of Osaka Bay is a broad expanse of land with a "zero-meter" elevation. The population is heavily concentrated in the coastal region, and various facilities and other property are also concentrated there. These factors make the coastal region fatally vulnerable to tsunamis. Are disaster mitigation countermeasures adequate to compensate for these vulnerabilities and minimize disaster? Here, too, many vulnerabilities exist in both hardware countermeasures and software countermeasures. These vulnerabilities are not unique to the Kinki region. They exist everywhere in the coastal regions of Japan.

As one vulnerability in hardware to tsunamis, first, on the southern coastal region, where attack by an 8 m tsunami is predicted, there are places where the height of the tsunami will exceed the crown height of the seawalls. In these places, a tsunami invasion cannot be avoided. Furthermore, many main highways and railway lines in this region have been constructed along the coastline. These lifelines may be cut by a tsunami, and this will hinder rescue work in a disaster and later recovery/reconstruction. In the inner bay area, the height of a tsunami will decrease in some cases, and the crown height of the seawalls is higher than the height of predicted tsunamis, but security is not guaranteed. Many of the coastal protection facilities in the coastal region are old, including the seawalls, and there is no guarantee that they can withstand a massive earthquake. If these seawalls are destroyed, they will lose their protective function in the tsunami that follows an earthquake. At present, it has been possible to confirm the earthquake resistance of only about 20% of these facilities, but surveys to confirm the earthquake resistance of the remaining 80% have not yet been conducted. In the inner bay, there is anxiety about whether water gates, land locks, etc. can be closed before a tsunami attack. There is also concern about reduced port functions, including delays due to a tsunami and damage to various types of facilities by drifting/impact by floating objects, etc. A variety of ships and boats converge in the bay and port, including large ships carrying hazardous materials, pleasure boats, fishing boats, and others. Therefore, there is concern that damage will not be limited to the ships themselves, but will include a chain of secondary and tertiary damage, such as damage to port and harbor facilities, degradation of port and harbor functions, and combinations of this damage.

As for delays in countermeasures by software, various problems are pointed out, including delays in the creation of a quick disaster prevention information communication system, disaster prevention education for the people of the region, disaster prevention training, etc. Preparation of hazard maps has also been delayed, even though these are indispensable both for improving these problems and enhancing disaster prevention awareness, and for evacuation during an actual tsunami attack. Hazard maps are prepared in city and town units. However, at present, only about 70% of all cities and towns in this region have prepared maps. In response to these delays, local governments (prefecture, city, and town governments) in the coastal region have specified items in connection with earthquakes and tsunamis in regional disaster prevention plans, and are working to lessen the predicted heavy damage. Nevertheless, it is a fact that there are large, fundamental problems in this regard. Each disaster prevention plan and set of tsunami countermeasures is limited to the administrative district of an individual local government, and there is no effective plan with compatibility for wide-area, regional cooperation exceeding local administrative divisions.

(2) *Establishment of a regional tsunami disaster mitigation plan in the Kinki coastal region*

In the Kinki coastal region, there are vulnerabilities in both hardware countermeasures and software countermeasures. Naturally, efforts by individual local governments are necessary in order to overcome these weaknesses. However, there is also a strong recognition of the necessity of efforts from the viewpoint of regional cooperation. Therefore, in 2005, related organizations, including the national government, local governments, and others, acting together, established an action plan which gives concrete form to basic policies for tsunami countermeasures. The aims of this action plan are to realize regional cooperative countermeasures and information sharing and to encourage autonomous action by the region. Based on this, there have been moves to establish tsunami disaster mitigation plans autonomously in the region itself.

The basic policies for tsunami countermeasures in the Kinki region and the action plan were established for the short-term (five years from the present) and mid- to long-term (20 years into the future). The target of the short-term policies is "to minimize human damage by tsunami," while the aim of the mid- to long-term policies is to "minimize damage by tsunamis, including not only human damage but also physical damage." Based on predictions of damage, which are a precondition for tsunami disaster mitigation countermeasures, and the roles which ports and harbors should play as nodes between land and sea, etc., the plan arranges information sharing and cooperation between concerned parties and indispensable countermeasures horizontally and summarizes the tsunami countermeasures to be taken jointly by the national government, local governments, and others with the aim of achieving the above-mentioned targets.

Regarding the short-term basic policies for the next five years, concrete urgent measures which should be implemented on a priority basis were studied cooperatively by the national government, local governments, and experts, and an action plan was established. Advance study of the network centering on ports and harbors by the national government and local governments is important for disaster mitigation. This must also include the "hinterland" (inland area served by a port). Recognizing this, the action plan clearly defines the division of roles among the national government, local governments, the private sector, and residents, and also requests voluntary participation by the people of the region.

First, the most important roles of the national government are a wide-area, regional response in terms of tsunami disaster mitigation countermeasures across prefectural borders and mutual coordination between disaster prevention organizations and others from the regional and national viewpoint. The roles of the national government also include technical support such as popularization of simple, step-by-step methods of diagnosing the earthquake resistance of structures, improvement in the speed and accuracy of tsunami information by practical application of GPS wave meters (GPS buoy system) deployed nationwide, etc. In addition, the national government provides financial support for the construction of disaster mitigation-related facilities by local governments.

Respective local governments set local targets based on the characteristics of the area under their jurisdiction of each local government. The action plan requires them to take comprehensive efforts, integrating software and hardware countermeasures, including preparation of tsunami hazard maps, automation/remote operation of water gates and land locks, reinforcement of facilities, etc. from the viewpoint of promoting concrete measures in the region.

From the viewpoint of the people of the region/disaster mitigation, the plan also touches on the importance of advance study, under normal conditions, of the network of ports and harbors, considering the hinterland, by the national government, local governments, and others.

Items to be implemented include (1) establishment of means of communicating information to crew members of ships while cargo is being handled or otherwise the ships are anchored at ports as well as to port workers and tourists, (2) improvement and sharing of tsunami information, (3) evaluation of the performance of facilities that have protective functions against tsunamis, (4) construction of such facilities and improvement of their functions, (5) countermeasures against washing away and drifting of cargo ships and small craft, (6) securing the functions of ports and harbors during disasters, and creation of disaster prevention bases and (7) improvement of their functions.

Regarding the residents and private sector, in order to further promote tsunami countermeasures, the cooperation of private-sector users of the ports is invited, and improvement of tsunami disaster prevention awareness through active participation of residents in disaster prevention education and practical disaster prevention training is requested. In the Kinki region, practical disaster prevention training has already been conducted in Gobo City (Hidaka Port), Wakayama Prefecture in 2005 and in Sakai City (Sakai Senhoku Port), Osaka Prefecture in 2006. This training, was conducted based on the participation of residents, assuming a tsunami attack caused by simultaneous Tonankai and Nankai earthquakes. Local governments, the Self-Defense Forces, and others also participated. Continuing promotion of these evacuation drills in various ports and harbors is also necessary. In addition, because the cooperation of persons who work in ports and have a good knowledge of the conditions around the port is indispensable if an actual evacuation is required, participation of such persons in disaster prevention drills is requested.

(3) *Tsunami disaster mitigation countermeasures led by local governments and the people of the area*

The above-mentioned action plan, specifying the basic policies and concrete implementation of tsunami countermeasures, was established by the national government and local governments working in cooperation.

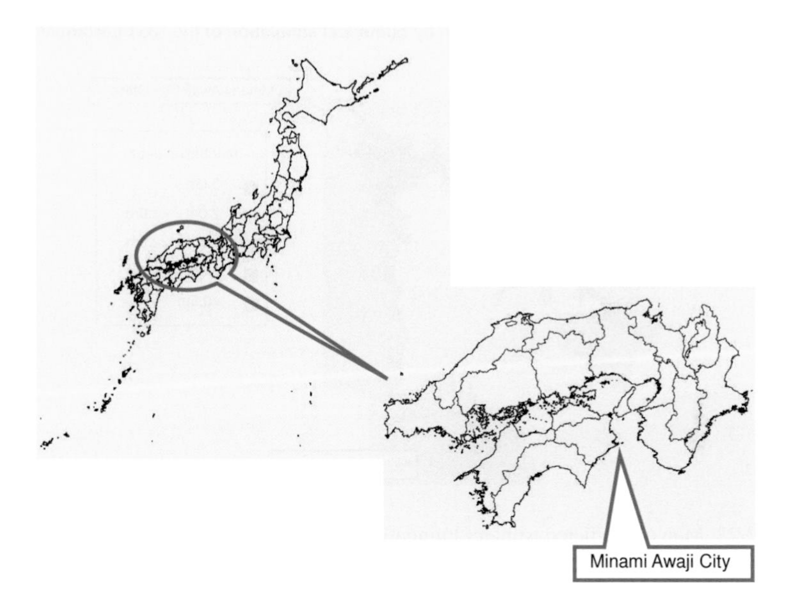

Fig. 3.21 Location of Minami Awaji City.

Based on this, there have been moves to establish tsunami disaster mitigation countermeasure plans voluntarily. As one example, the following will describe the Tsunami Countermeasures Study Council in Minami Awaji City under the leadership of Hyogo Prefecture (Fig. 3.21).

Minami Awaji City has a population of 54,000 persons, which is concentrated in the coastal area, centering on Fukura Port. In the past, Minami Awaji suffered attacks by large tsunamis in 1854 and 1946. Fortunately, The death toll was zero in the tsunami caused by the 1946 Showa Nankai Earthquake (*M*8.0). However, during the 60 years since that tsunami, the city has become increasingly more vulnerable to earthquakes and tsunamis, as the population has aged and houses and other buildings have deteriorated over time. Recently, there is concern that the region's preparedness has decreased. Therefore, assuming the 1854 earthquake (*M*8.4) and the 1946 earthquake (*M*8.0) occurred today, the tsunamis caused by those earthquakes were reproduced by numerical simulation. According to the results (Fig. 3.22), after a *M*8.4 earthquake,

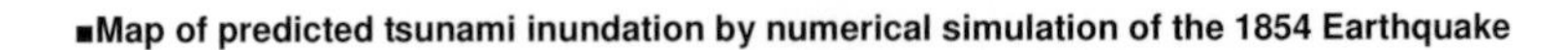

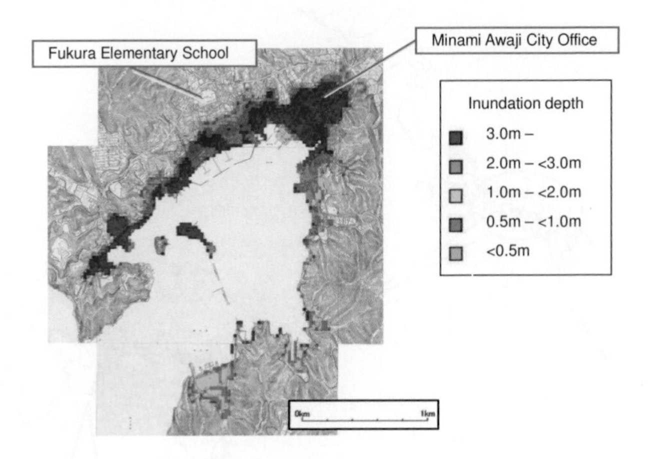

Fig. 3.22 Map of predicted tsunami inundation (source: Hyogo Prefecture homepage).

a tsunami with a maximum height of 5.8 m would attack Fukura Port in Minami Awaji City. In contrast to this, the crown height of the seawall which is being constructed in Fukura Port as defense against storm surges in typhoons is only 2.2-2.95 m. A drop ("subsidence") of approximately 40 cm in the neighboring ground is considered possible in the earthquake that causes the tsunami. If this happens, the crown height during the tsunami attack would be 1.8-2.55 m, or 3-4 m lower than the height of the tsunami. A detailed study was made of the 33 hectares of densely populated land behind Fukura Port (Fig. 3.23), and the results of the investigation of predicted damage were reported. In the district for which damage was predicted, the estimated death toll in a *M*8.0 scale earthquake would be approximately 60 persons, and property damage would reach ¥12 billion, or more than US$110 million (exchange rate: ¥108 = US$1). In a *M*8.4 earthquake, the death toll and property damage would rise to approximately 90 persons and ¥27 billion (well over US$200 million), respectively. Based on this, Minami Awaji City decided to implement tsunami disaster mitigation countermeasures aiming at a protection level of zero deaths in a *M*8.0-scale earthquake tsunami as a

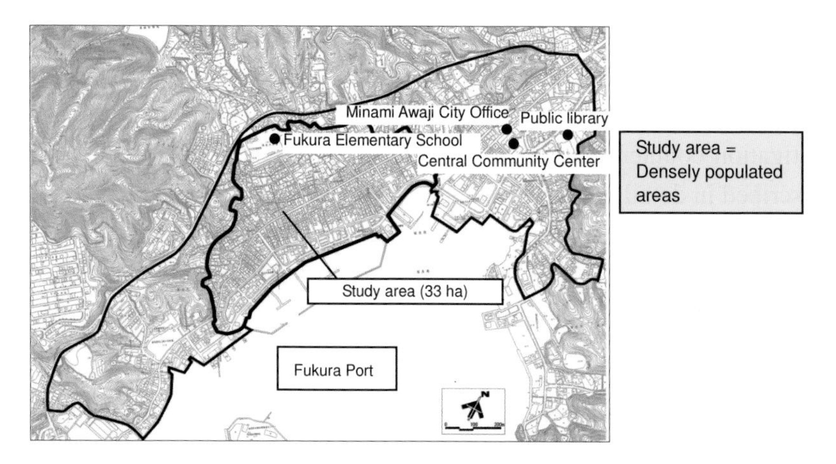

Fig. 3.23 Area damage prediction study area (source: materials of Study Council).

five-year target and a disaster mitigation level of zero deaths and minimal economic impact in a *M*8.4-scale earthquake tsunami as a 10-year target.

The Tsunami Countermeasures Study Council consisted of the following members: included representatives of the city's neighborhood associations and women's societies, which are central to various activities of the region, representatives of the Fire Department, which is the front line in regional disaster prevention, representatives of the fisheries cooperative, who have a detailed knowledge of conditions in the neighboring waters, persons in charge in related government organizations (national, prefecture, and city levels), and experts with knowledge of tsunamis and tsunami disasters. Because many local people with a knowledge of the region participated in the Study Council, the people of the region felt that the countermeasures were their own work. As a result, they can expect active action by residents during an actual tsunami attack. In this Study Council, the city studied concrete measures for expected damage conditions (human damage and damage to houses and other buildings) based on a tsunami hazard map prepared by Minami Awaji City, and clarified the division of roles for each tsunami disaster mitigation measure (person or group responsible for implementation) and the time period for implementation of countermeasures. The Study Council recommended tsunami disaster mitigation countermeasures consisting of six items. The main items are described in detail in the following:

(1) *"Hardware" infrastructure aiming at zero deaths*

Because there are limits to the time and financial resources that can be invested in hardware infrastructure for tsunami countermeasures, it was recommended to start early from projects which would be completed within practical limits utilizing the existing stock as much as possible. For example, hardware measures to be implemented in order to achieve zero deaths in a tsunami due to a *M*8.0 earthquake (assuming the 1946 Showa Nankai Earthquake) in a timeframe of five years included extension and reinforcement of existing seawalls, which is also effective against tsunamis, and construction of a tsunami disaster prevention station to perform remote monitoring/operation for closing the gates at openings in the seawalls during a tsunami attack.

A larger tsunami following a *M*8.4-scale earthquake (assuming the 1854 Ansei Nankai Earthquake) is expected to cause greater damage. However, it was recommended that hardware measures against a tsunami of this scale should be carried out prudently. The reason for this careful policy is as follows: In order to reduce damage by hardware measures alone, it would be necessary to construct sea walls as high as looking up in order to reduce invasion of the bay by the tsunami. This would require a huge investment of money and time, and as negative impacts, the scenic beauty of the area would be degraded. Considering these problems, it was recommended to implement hardware measures which are relatively less expensive and can be completed in a shorter time, such as construction of high ground as a tsunami refuge area, reinforcement of buildings, etc., in combination with software measures, such as holding workshops for residents, cooperation with school education, etc.

(2) *Construction of a tsunami disaster prevention station*

A seawall is being constructed in Minami Awaji in order to prevent invasion of the land by sea water during a tsunami or storm surge. ("Storm surge" means high sea level caused by a typhoon, etc. The Hurricane Katrina disaster in the US was caused by storm surge breaching the levees in New Orleans.) Basically, a seawall only demonstrates this function when it is constructed continuously along the coastline. However, openings are provided at various places along a seawall to give people easy access to the ocean under normal conditions. During a tsunami attack, these openings are closed by tsunami control gate. A tsunami disaster prevention station is a facility to be constructed in order to promptly close the water gates and other openings in seawalls prior to a tsunami attack. Because this is a control center, the city had planned to construct it on a mountaintop to avoid any possibility of inundation damage to the station itself. However, the Study Committee later changed the plan and proposed to construct the station beside the sea. This was the result of a conclusion that this should be a facility with compound functions. In addition to the basic function of opening and closing the land locks, these include a function of communicating tsunami information, a function as an emergency place of refuge, and a function of resident education and publicity. The following are concrete examples of individual opinions.

- The experts pointed out that people lose their footing and cannot walk in a tsunami with an inundation depth of only 30 cm. For this reason, evacuation from the coast to a high location inland becomes increasingly difficult when people are closer to the shore or delay the start of evacuation. Therefore, a place that enables "vertical evacuation" is necessary near the coast.
- People easily panic in a major disaster. Considering the psychology of residents, people will tend to flee to a familiar facility that they use regularly, rather than to an unfamiliar facility. Therefore, a emergency place of refuge should be a place that people see every day.

- For tourists and other visitors who do not know the evacuation routes to inland areas, evacuation to a facility in an easy-to-see location is simpler.
- In order to increase visibility to the people of the area and help heighten tsunami disaster prevention awareness by use as a facility for elementary school field trips to study disaster prevention and for regular exhibitions on disaster prevention, an easily-accessible facility located at the coast is more effective.
- If the station is near land locks and other tide control facilities, it will be possible for station personnel to respond to changing unexpected conditions in cooperation with the person in charge at the site.

Based on a total consideration of these above ideas, and also intending to secure a place of refuge which has adequate earthquake resistance and will not be inundated in a tsunami, they changed the initial plan and proposed to construct the tsunami disaster prevention station near the sightseeing boat boarding dock. This is a place that residents see every day, and also a place where tourists gather.

(3) *Cooperation with school education*

Improving the disaster prevention awareness of the area's children through education in school also has the spreading effect of heightening the disaster prevention awareness of the people of the area as a whole. For full-scale efforts in this type of school education, first, workshops on methods of disaster prevention education and the proper form of educational materials were held for teachers in elementary and middle schools.

(4) *Establishment of a plan for "disaster prevention town-building"*

The establishment of a disaster prevention town-building plan, which is the foundation for constructing a "disaster prevention culture" of "self help," "mutual help," and "public help," is important and also tireless efforts through cooperation between the people of the region and government authorities based on that plan, are necessary. To give only two examples of "disaster prevention town-building," the government raises the ground level when it replaces public facilities, and the government provides examples to residents when they build new houses, and residents strengthen the disaster prevention capabilities of the town as a whole by remembering these disaster prevention countermeasures when constructing a new house or remodeling an existing house.

Advance study of plans for recovery/reconstruction after a major disaster is extremely important for quick and appropriate recovery/reconstruction work after a disaster actually occurs. This type of study is also important for ensuring that residents understand the seriousness of a tsunami disaster. Measures for reconstructing the town after a future tsunami attack may include condemning land and raising the elevation of residential ground. Taking the time to plan these countermeasures in advance is important for leaving a disaster-resistant town to the coming generations.

Part II

Tsunami Behavior and Forecasting

Chapter 4

Occurrence and Amplification of Tsunamis

4.1 Mechanism of Tsunami Occurrence

(1) *Earthquakes which cause tsunamis*

At the time of the Meiji Sanriku Earthquake Tsunami (1896), the mechanism by which earthquakes cause tsunamis was not well understood and several theories had been advanced. Typical examples are the seaquake theory and the seafloor deformation theory. According to the seaquake theory, when the period of oscillation in an earthquake coincides with the natural period of a bay, the water level in the bay undergoes large deformations as a result of resonance, and this becomes a tsunami. Because this theory seems to explain the experience that tsunamis attack after strong tremors, it seems to have been generally accepted at the time. However, this theory does not adequately explain the following facts: (1) tsunamis attack various areas after an earthquake, including areas which are not bays, (2) tsunamis sometimes attack after a certain time is elapsed since an earthquake, and (3) tsunamis frequently attack from far outside a bay, rather than inside the bay. On the other hand, the seafloor deformation theory hold that a deformation of the seafloor during an earthquake causes displacement of the ocean surface directly above that deformation, and this is transmitted as a tsunami. The seafloor deformation theory satisfactorily explains the phenomena that could not be explained by the seaquake theory. After several controversies, around the beginning of the 20th century, the seafloor deformation theory was recognized as the general mechanism by which tsunamis occur, based on various research results such as the one that shows the enormous energy of tsunamis is larger than could be explained by the seaquake theory and corresponds to the energy predicted by the seafloor deformation theory. However, if the phenomenon of natural oscillation, which forms the basis for the seaquake theory, is understood as resonance due to coincidence of the period of the tsunami and the natural period of a bay, this concept remains valid as an explanation of the mechanism of amplification of the wave height of tsunamis.

When the theory of seafloor deformation by earthquakes became clear, the relationship between earthquakes and the occurrence of tsunamis could also be understood. When an earthquake occurs, causing a sudden shift in the ground, tremors called seismic ground motion are generated and transmitted through the earth's crust as seismic waves. At this time, a displacement called a fault occurs at the site of this shifting motion. The effect of this displacement also appears at the surface of the affected ground. In general, displacement at the ground surface increases as the scale (magnitude M) of an earthquake increases or the depth of the focus of the earthquake becomes shallower. When an earthquake occurs at the sea bottom, the displacement at the ground surface disturbs the sea water above that ground. This is the initiation of a tsunami. Therefore, the scale of a tsunami increases as the magnitude of the earthquake increases and the depth of the focus of the earthquake becomes shallower. The following section will explain sea bottom ground deformations more in detail using a fault model.

(2) *Relationship between displacement of the sea bottom ground and tsunami*

Seismic energy (or moment) is evaluated by the product of the area of the fault and the amount of dislocation at the focus. Therefore, assuming the fault area is constant, earthquakes with larger energy will produce larger amounts of dislocation (i.e., displacement). In this, the area of the fault (length x width) corresponds to the special expanse of the tsunami source, and the amount of dislocation is converted to amounts of vertical and horizontal displacement, which are related to the amplitude of the tsunami, considering the direction of the dislocation (slip) and the inclination of the fault (dip). However, because the displacement becomes

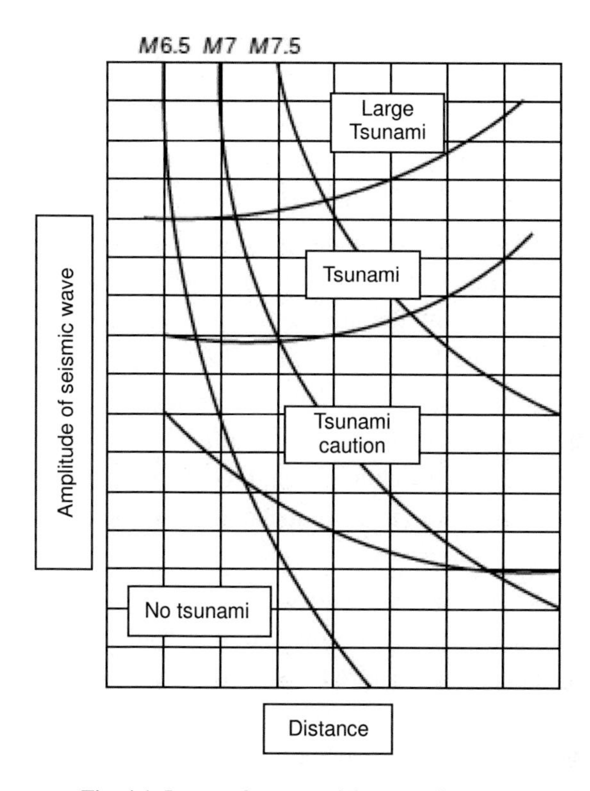

Fig. 4.1 Image of a tsunami forecast diagram.

smaller three-dimensionally with distance from the focus, the displacement at the sea bottom surface decreases as the depth of the focus increases. This means the amplitude of the tsunami also becomes smaller. Conversely, the displacement of the sea bottom increase when the focus is shallower.

Such relationship between earthquakes and tsunamis is used in the tsunami forecast diagram (Fig. 4.1), which has been employed in tsunami forecasting since around 1941. This is an empirical diagram, which is based on the relationship between the magnitude of past earthquakes and the scale of the tsunamis caused by those earthquakes, and assumes that (1) the scale of a tsunami is determined by the magnitude of an earthquake and (2) attenuation due to the spread of a tsunami is determined by distance from the focus of the earthquake. When an earthquake is observed by a seismograph at a certain location, the amplitude of the seismic wave (corresponding to the *y*-axis in Fig. 4.1) is read immediately from the recording paper. The time to the arrival of the primary wave (P wave) and time to arrival of the secondary wave (S wave) are also read, and the distance from the focus of the earthquake (*x*-axis in Fig. 4.1) is estimated from the time difference between the two; in other words, from the duration of preliminary tremors caused by the P wave before the S wave arrives. The scale of the tsunami which will attack at the location where the earthquake was observed is estimated from these two values and Fig. 4.1. Data plotted near the lower left corner indicate "no tsunami." From this diagram, it is also possible to estimate the magnitude of an earthquake. For example, if the magnitude is estimated to be *M*7, a "large tsunami" forecast is necessary for the coast near the focus. As the distance from the focus increases (the right side on the *x*-axis), a "tsunami" forecast or a "tsunami caution" forecast is necessary. Thus, the required forecast is derived from the actual distance from the focus of the earthquake to the coast. Strictly speaking, however, as mentioned above, what determines the scale of a tsunami is not the distance from the focus of an earthquake to the coast, but the distance to the ocean floor immediately above the focus (in other words, the depth of the focus). In current quantitative forecasts, the effect of depth is effectively introduced because a fault model is used, as described below.

The cause of tsunamis can be understood to a certain extent by the seafloor deformation model. However, for quantitative handling of seafloor deformations, which are the cause of tsunamis, it is necessary to model the fault motion of the earthquake. In the 1960s, a fault model assuming a rectangular fault plane was proposed as the mechanism of tsunami initiation, enabling quantitative formulation of the conditions for occurrence of tsunamis (initial source of tsunami). This model is composed of six static parameters, which express geometric features (see Fig. 4.2), and two dynamic parameters, which express kinematic characteristics (rupture propagation and time from start to completion of

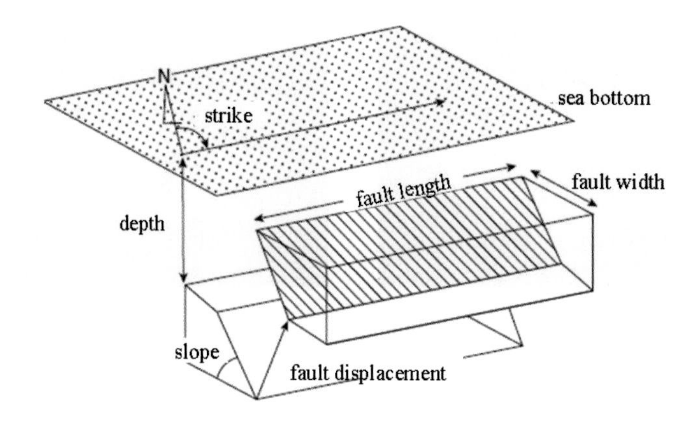

Fig. 4.2 Earthquake fault model.

displacement after rupture propagation). The final amount of sea bottom displacement caused by fault motion is expressed by the former parameters. When actually reproducing a tsunami by numerical simulation, the calculation is performed under initial conditions assuming the vertical component of the final sea bottom displacement is directly reflected in the displacement of the sea surface.

Figure 4.3 shows the sea bottom displacement frequently seen in plate boundary earthquakes and the resulting change in water level. The figures show the condition 1 min 4 s (Fig. 4.3(a)) and 2 min 54 s (Fig. 4.3(b)) after the earthquake, respectively. In both figures, the sea bottom topography is assumed as a horizontal plane. However, in actuality, the right side of these figures leads to a coastal area with a shallow water depth, and the left is deep sea. The coastline is located where the water surface forms a jagged line. Together with the occurrence of a fault, the sea bottom ground begins to rise in some part and sink in other part, and accompanying this, the same changes take place at the sea surface (Fig. 4.3(a)). This is the initiation of a tsunami. After the fault motion is complete, the changes in the sea surface begin to propagate in all directions (Fig. 4.3(b)).

Figure 4.4 shows the distribution of sea bottom displacement during the Solomon Island Earthquake, which was a typical trench-type

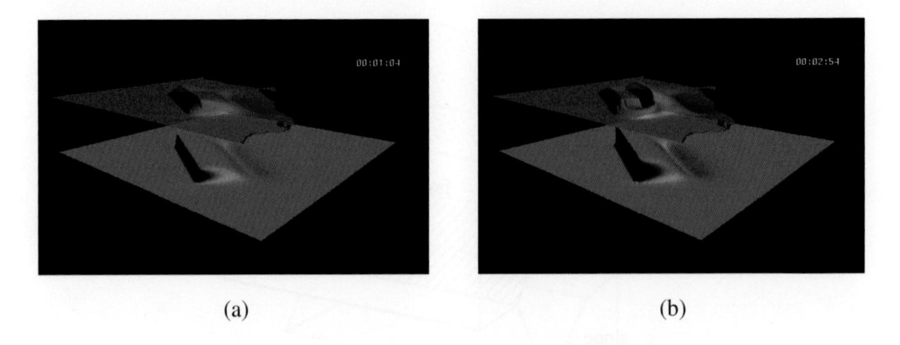

Fig. 4.3 Initiation of a tsunami by seafloor deformation.

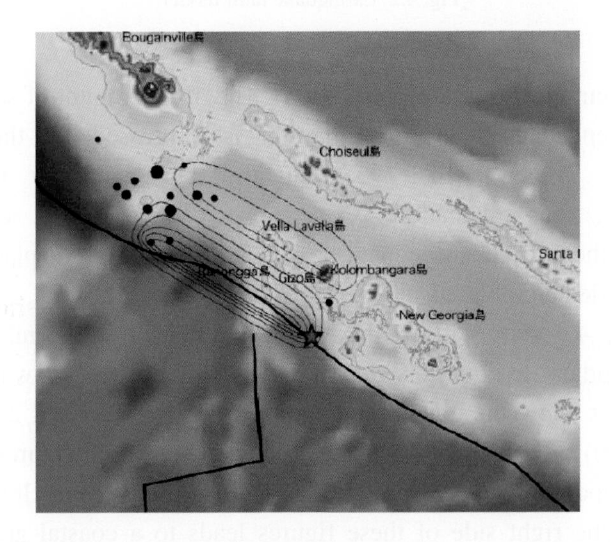

Fig. 4.4 Displacement of the sea bottom by the fault in the Solomon Island Earthquake which occurred in April 2007 (drawn by Koshimura of Tohoku University).

earthquake, using displacement contour lines with a 50-cm interval. The solid lines show upheaval, and the broken lines show subsidence. The sea bottom rose due to rebound and uplift of the landside plate (lower left side in the figure). The upheaval area was approximately 200 km in major diameter, and the maximum amount of upheaval was 2.9 m. The subsidence area is to the upper right of the upheaval area. Maximum subsidence was 1.3 m. This displacement of the sea bottom became the initial source of a tsunami in the same form that it occurred at the seafloor. As a result, the water level dropped in the subsidence area, and a tsunami attack on the islands at the upper right began with a receding backwash current. On the other hand, because the sea surface rose in the upheaval area, the first motion of the tsunami to the lower left direction was a rising wave. In other words, the first-motion characteristics of a tsunami depend on the condition of displacement at the sea bottom. The first motion of a tsunami may be a receding backwash current at the coast, or a rising wave depending on the direction of propagation. However, it is not possible to predict in advance precisely what type of sea bottom displacement will occur. Therefore, the conventional wisdom "tsunamis always begin with a suddenly receding tide" is not necessarily true.

Normally, a faulting event ends several seconds to several minutes after motion begins. Duration on this order can be considered instantaneous for tsunamis, which have a very long period. However, when a duration is larger than the several minutes, the effect is manifested in the earthquake tremors. As a result, the short period component decreases and the long period component becomes predominant. The maximum seismic amplitude of earthquakes is governed by short period waves. Therefore, the maximum amplitude tends to decrease, resulting in underestimation of the magnitude of the earthquake. However, because the fault moves over several minutes, the final displacement of the sea bottom is large, resulting in a large tsunami. In other words, even if the magnitude of an earthquake is small, it may produce a large tsunami. This kind of earthquake is called a "tsunami earthquake." If the motion is even slower, another type of effect appears in the tsunami. The tsunami component generated by the initial displacement of the sea bottom begins to propagate while the earthquake is still in progress, and has already moved a considerable distance when the earthquake motion ends. This disperses the wave height of the tsunami, and thereby decreases the height of the tsunami.

Reference

Imamura, F. (2001): Tsunami damage and simulation, Transactions of the Japan Society of Computational Engineering and Science, Vol. 6, No. 3, pp. 311-315. (in Japanese)

4.2 Amplification and Attenuation of Tsunamis at Coast

(1) *Propagation attenuation*

Because the propagation area increases as a tsunami propagates in all directions to greater distances from its source, the energy of the tsunami is dispersed and attenuated. Under the conditions that the area of the tsunami source area is small, that is, a point source, and the water depth is uniform, a tsunami is transmitted in a concentric circular shape, as illustrated in Fig. 4.5, and the height of the tsunami is attenuated in inverse proportion to the square root of the distance from the center of the tsunami source. However, actual tsunami source areas are large and are elliptical in shape, with a length of several 100 km and a width of several 10 km. Therefore, the degree of attenuation of the tsunami wave

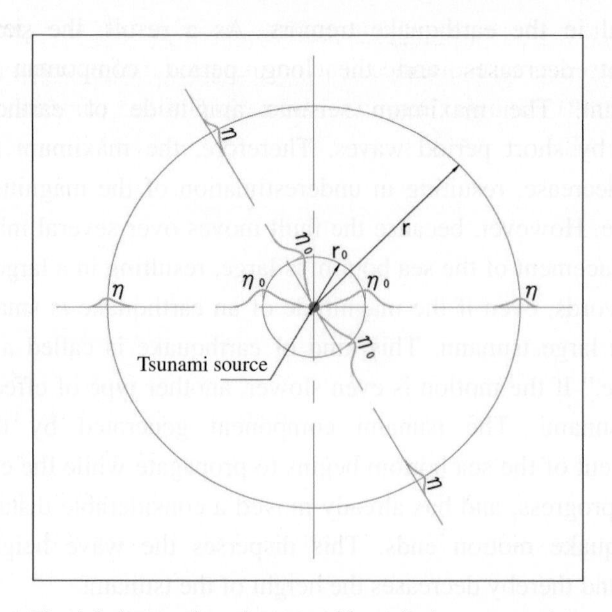

Fig. 4.5 Attenuation of a tsunami by radiating propagation.

height will differ depending not only on the distance from its source area, but also on the direction of propagation. In the Great Indian Ocean Tsunami on December 26, 2004, the tsunami source area was approximately 1,000 km in length from north to south. Large tsunami attacks occurred in areas of Sri Lanka, India, and the Phuket coast in Thailand, which face the lengthwise axis, while only a small tsunami propagated to Myanmar and Bangladesh, which lie along the extension of this lengthwise axis. Furthermore, as will be described below, deformation of a tsunami also includes the effects of shoaling effect that occurs as the water depth changes, and refraction, in which the tsunami wave is “bent” by changes in the topography of the sea bottom.

In the process of propagation, the height of a tsunami is also reduced by energy loss resulting from the viscosity of the water and friction with the sea bottom. However, the reduction of tsunamis by these factors is small. Unless a tsunami propagates over a long distance of more than 1,000 km, the effect of these factors does not become apparent. Because energy loss is large in the shorter period component of the tsunami, the energy loss of a tsunami due to propagation over long distances is also greater in the short period component. As a result, the period of a tsunami increases with distance. The normal period of a tsunami is on the order of 10-40 minutes. However, the period of the tsunami caused by the Great Chilean Earthquake of 1960, which reached Japan 18,000 km away after traveling for roughly 24 hours across the Pacific Ocean, was more than one hour.

(2) *Shoaling effect*

A tsunami caused by an earthquake that occurred in deep water near an ocean trench rapidly increases in size when it reaches the shallow coastal waters. This change in the size of a tsunami as the water becomes shallower is called shoaling effect. The reason why such phenomenon is caused is explained as follows: Considering the continuity of tsunami energy flux, because the speed at which a tsunami is transmitted is reduced by the shallower waters (actually, this means the propagation

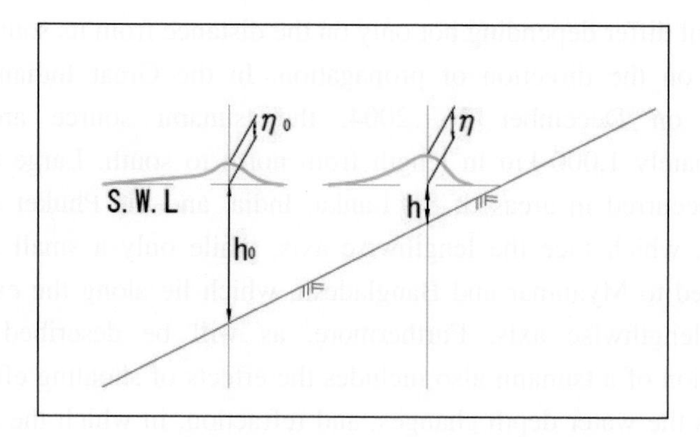

Fig. 4.6 Shallow water transformation of a tsunami.

velocity of the energy of the tsunami; however, due to the long period of tsunamis, this is the same as the speed of transmission of the tsunami waveform), the energy of the tsunami must increase. Based on this, the height of a tsunami changes as shown by the following equation:

$$\eta/\eta_0 = (h/h_0)^{-1/4} \tag{6}$$

where, η and η_0 are the heights of the tsunami at the water depths h and h_0, respectively, as shown in Fig. 4.6. The subscript 0 represents an arbitrary point offshore. For example, if a tsunami is generated in water with a depth of 4,000 m and transmitted to a point where the water depth is 10 m, $\eta/\eta_0 = (10/4000)^{-1/4} = 4.47$. This means that the height of the tsunami is amplified by approximately 4.5 times.

(3) *Refraction*

The speed at which a tsunami is transmitted is given by Eq. (1). As can be understood from this equation, the propagation velocity of a tsunami increases with water depth. When the wave celerity changes due to water depth in this manner, the direction of advance of a wave is changed by the sea bottom topography. This phenomenon is called refraction.

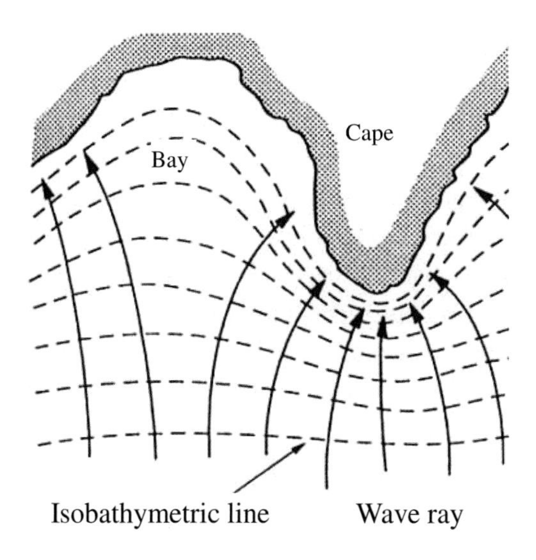

Fig. 4.7 Refraction of a tsunami.

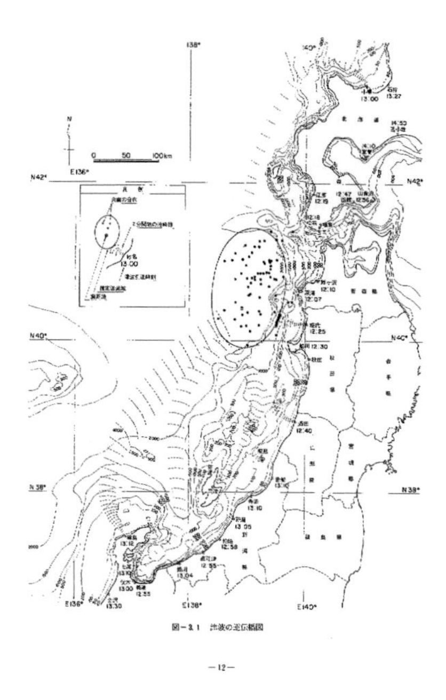

Fig. 4.8 Determination of the tsunami source area by inverse-propagation calculation of a tsunami.

Due to refraction, a tsunami is increasingly bent toward shallower water, where its wave celerity is slowed. As a result, the wave rays tend to cluster at coasts where the isobathymetric lines (equal depth lines) jut out toward offshore, for example, at the tip of a cape, as shown in Fig. 4.7. Consequently, the energy of the tsunami concentrates in this area, and the height of the tsunami wave increases. On the other hand, at a coast where the isobathymetric lines are caved in toward land such as in a bay, the wave rays diverge and the energy of the tsunami is dispersed. It might be thought that the height of the tsunami wave will decrease here. Actually, however, the size of the tsunami will increase at the back of the bay. The reason for this increase is the phenomenon of reflection, which is discussed in the following section.

The arrival time of the first wave of a tsunami can be known by observing the tsunami with tide gauges located at various points on a coast. Because the wave celerity of a tsunami is determined by the water depth, it is possible to trace the route that the tsunami took to arrive at various locations back to its source if the arrival time is known. This is called inverse-propagation analysis of a tsunami. If the tsunami is traced back to the time when the earthquake occurred, the point where the propagation began can be located. This means that the tsunami source area can be estimated by inverse-propagation analysis from the various points where the tide (tsunami) level was observed. Figure 4.8 is a diagram in which the source area of the 1983 Nihonkai-Chubu Earthquake Tsunami was estimated by inverse-propagation analysis. Because it was not possible to obtain tide level records from Russia, China, or Korea, the seaward side boundary of the source area is unclear. The displacement of the sea bottom ground can be determined with higher accuracy by comparing the tsunami source area estimated by this method and the changes in sea bottom topography obtained from the earthquake ground motion.

(4) *Reflection*

It is often said that care is necessary in V-shaped bays because the height of the tsunami wave increases due to concentration of the tsunami energy in the back of the bay. The height of tsunamis increases in the back of

V-shaped bays because the tsunami wave is reflected by the two sides of the bay, and this tsunami then advances with refracting into the narrow back of the bay, where its energy concentrates. Because the wave length of a typical tsunami is several 10 km and its height is extremely small in comparison with its length, tsunamis are reflected by coasts where ordinary waves simply break and are not reflected. Although a reflected tsunami initially heads back toward offshore, as shown in Fig. 4.9, it is refracted toward shallower waters and, as a result, returns to the coast. A tsunami advances into the back of a bay while repeatedly undergoing reflection and refraction. Finally, the energy of the tsunami concentrates at the back of the bay, and the size of the tsunami increases. In contrast, ordinary wind waves break on the two coasts of a bay and almost all of their wave energy is lost; hence, they virtually never form reflected waves. For this reason, the wave height of wind waves does not increase in the back of a bay.

Because tsunamis are reflected by any coasts, it is generally thought that the tsunamis that occur in the Sea of Japan, which is surrounded by land, undergo very slow attenuation, as they are reflected by the land numerous times. There are also cases in which large tsunamis occur due to unanticipated concentration of reflected tsunami waves.

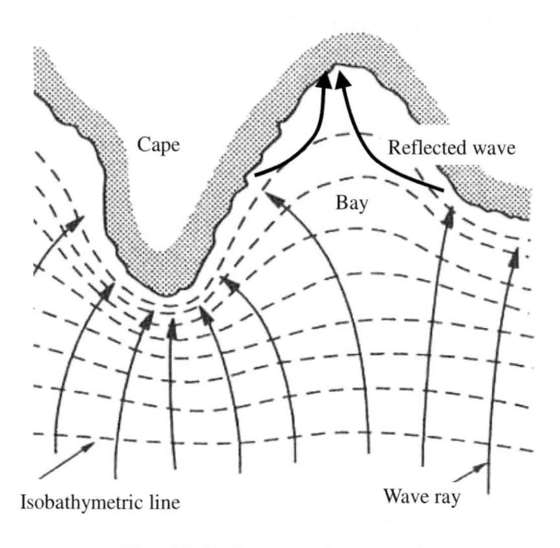

Fig. 4.9 Reflection of a tsunami.

(5) *Diffraction*

Tsunami waves have the same properties as ordinary wind waves, only having a longer period in comparison with ordinary waves. Therefore, if the advance of a tsunami is obstructed by a structure, or by a geographical feature such as an island, peninsula, or cape, it will circle around to the back of that obstruction (Fig. 4.10). This phenomenon is called diffraction.

However, because the period of a tsunami is long, its wavelength is also long. The wavelength of a wind wave is on the order of 100 m. In contrast, a tsunami with a period of 15 minutes has a wavelength of 9 km in 10 m deep water. At an island with a size of around 50 m, the size of the island is very small relative to the wavelength of the tsunami, and the change in the size of the tsunami due to diffraction will not be apparent. Changes in the size of a tsunami due to the diffraction phenomenon become remarkable when the scale of an island, cape, or peninsula exceeds 0.5 km. When the shielding effect of the topography of these geographical features is large, the tsunami that circles to the back of a cape, island, or peninsula will become small. However, care is necessary, because there are cases in which the size of a tsunami increases due to a superposition of the waves reflected from land.

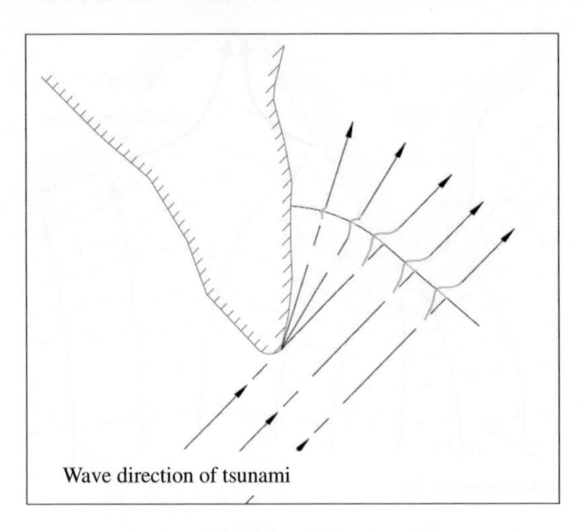

Fig. 4.10 Diffraction of a tsunami.

As mentioned previously, in the case of tsunamis, the phenomenon of refraction occurs at all water depths on the planet. Therefore, diffraction and refraction may occur simultaneously. Which of these phenomena is predominant will depend on topographical conditions. This means that the size of a tsunami will not necessarily decease at the back of a cape.

(6) *Resonance*

Oscillation sea water in a bay has a natural period that corresponds to the shape of the bay and the distribution of water depth. The natural period includes many modes, from low-order modes with long periods to high-order modes with short periods. When a tsunami which is close to the natural period of a bay invades the bay from the open sea, it displays large oscillation in a mode corresponding to the natural period of the bay. This phenomenon is called resonance. Many natural periods exist. However, the natural periods at which the phenomenon of resonance occurs are limited to oscillation of low-order modes.

Figure 4.11 shows the amplification of a tsunami in a bay caused by resonance. The y-axis in the figure is the ratio of the height of the tsunami at the back of the bay to the height of the tsunami at the mouth of the bay H/H_0. The size of a tsunami in a bay increases as this ratio increases. The x-axis shows the natural period of oscillation of sea water in the bay T. The white squares are results from the 1933 Great Sanriku Tsunami; the black squares show the 1960 Chilean Earthquake Tsunami. The 1933 Great Sanriku Tsunami is a tsunami that occurred off the Sanriku Coast of northeastern Japan. The period of a tsunami that occurs in waters close to the coast is comparatively short, being from 10 to 30 minutes. On the other hand, the Chilean Earthquake Tsunami occurred off the coast of Chile, on the opposite side of the Pacific Ocean from Japan. Traveling at the speed of a jet plane, this tsunami reached Japan in roughly one day and caused heavy damage on the Pacific coast. Because this tsunami reached Japan by long-distance propagation crossing the Pacific Ocean, the short period wave

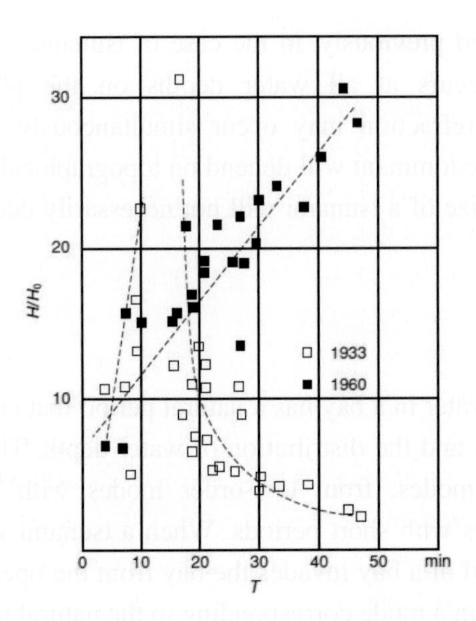

Fig. 4.11 Amplification of tsunamis in bays by resonance. 1933: Sanriku Tsunami; 1960: Chilean Earthquake Tsunami.

component had attenuated during propagation. As a result, the period of the wave was extremely long, being on the order of one hour.

According to Fig. 4.11, in the 1933 Great Sanriku Tsunami, the tsunami was amplified greatly in bays when the natural period of the bay was closer to 15 minutes. Based on this, it can be said that the tsunami was greatly amplified by resonance in bays with natural periods closer to 15 minutes because the period of the tsunami was roughly 15 minutes. Although no data are available for bays with natural periods exceeding one hour, in the Chilean Earthquake Tsunami, within the range for which data are available, the amplitude of the tsunami increased in bays as the natural period of bays increased. Thus, even in the same bay, the degree of amplification will differ, depending on the period of the attacking tsunami. In cases in which the period of the incident tsunami is close to the natural period of a bay, there is a high possibility that the tsunami will be amplified by the phenomenon of resonance. Therefore, it is also necessary to consider the period of tsunamis.

(7) *Trapping of tsunamis*

When a tsunami attacks an island or cape, or the continental shelf, it is reflected by the coast and returns toward the open sea. However, waves with a designated period corresponding to the sea bottom topography return toward the land due to deepening of the water. Then, it is reflected by the land again, but eventually returns back to the land once again. Sometimes, this process is repeated. In the case of islands, as shown in Fig. 4.12, the tsunami will be trapped by the area surrounding the island, and will be unable to escape from this area. This phenomenon is called trapping. At islands, the phenomenon of trapping occurs on the coast opposite the side where the tsunami directly attacks, and large tsunamis have been observed in some cases. For example, in the Great Indian Ocean Tsunami on December 26, 2004, a large tsunami run-up height was observed in Sri Lanka in an area that was shielded from the tsunami. It is thought that this may have been the result of the trapping phenomenon.

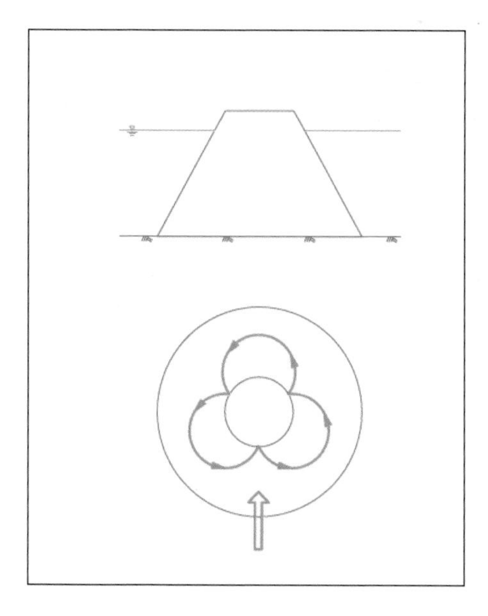

Fig. 4.12 Wave trapped around an island.

The tsunami trapping phenomenon also occurs on the continental shelf. Although the incident tsunami is reflected by the land and returns toward the sea, it is reflected back to the land because the water depth on the sea side increases suddenly at the edge of the continental shelf. When this process is repeated, the tsunami cannot escape from the continental shelf. The waves that are trapped in this way are those of a designated period that is determined as a result of the topography such as the length of the continental shelf. If the period of the tsunami wave that becomes trapped is close to the average period of the tsunami, large tsunami energy will be trapped. Waves which are trapped in the coastal area are also called edge waves.

Because the trapping phenomenon is a phenomenon that increases sharply near the shore line, the effect of sea bottom friction in shallow waters is strong, and the tsunami energy is easily lost. Therefore, although the trapping phenomenon is caused by a designated component of the tsunami period, the so-called strong resonance phenomenon tends not to occur.

4.3 Tsunami Run-up and Wave Force

(1) *Tsunami run-up height*

In general, the leading edge of sea water moves when an incident wave approaches a coastal area. When the leading edge of the sea water comes into contact with a vertical seawall, the wave front moves vertically up and down. On the other hand, at sloping embankments and natural beaches, the wave front moves as though running up on a beach. This is called the "run-up" of a wave. The maximum height which the wave front reaches is called the "run-up height." The same phenomenon also occurs in the case of tsunami waves. However, the run-up of ordinary waves is completed in several seconds; in contrast, because tsunami waves have a long period, run-up continues for several minutes.

Excluding certain slightly exceptional phenomena, tsunami run-up height is an extremely useful parameter for expressing the size of tsunamis. These slightly exceptional phenomena include, for example, cases in which the tsunami passes over a sloping beach, invades a flat hinterland, and then continues to surge inland for more than 10 minutes. In such cases, the final position where the wave front stops is not necessarily the location with the highest run-up. As another exception, in the earthquake focal area, the incident tsunami does not approach from the sea, but rather, radiates from the coast toward the open sea. In this case, such run-up that corresponds to the scale of the tsunami does not occur.

Information on tsunami run-up height is directly useful when considering refuge areas. Either of the following is an appropriate place of refuge in a tsunami:

(1) Outside the inundation area (run-up position): Horizontal evacuation;
(2) Higher than the run-up height (strictly speaking, the inundation height at a given location): Vertical evacuation.

If a place which is higher than the run-up height exists, it is often faster to evacuate to that place rather than attempting to evacuate to a location outside the inundation area. If persons evacuate to a location higher than the run-up height before the tsunami attack, they will not be swept away by the tsunami. Thus, run-up height is useful information for judging refuge areas during a tsunami attack.

As the most fundamental case, here, let us consider run-up of a tsunami when a uniformly sloping section is connected to one-dimensional horizontal floor. The linear theoretical solution for the run-up height R when a tsunami with wave height H (when the amplitude is a, $H = 2a$) is incident on a beach with this cross section is as follows (Shuto, 1972):

$$\frac{R}{H} = [(J_0(4\pi\ell / L))^2 + (J_1(4\pi\ell / L))^2]^{-1/2} \tag{7}$$

where ℓ is the horizontal length of the sloping section in water and L is the wavelength of the tsunami on the horizontal floor section. $J_0(4\pi\ell / L)$ and $J_1(4\pi\ell / L)$ are Bessel functions. The following approximations are applicable when $4\pi\ell / L$ is sufficiently large or small:

$$\frac{R}{H} = \sqrt{2\pi}\sqrt{\ell / L} \quad \text{(when } 4\pi\ell/L \text{ is large)} \tag{8}$$

$$\frac{R}{H} = 1 + 2\pi^2(\ell / L)^2 \quad \text{(when } 4\pi\ell/L \text{ is small)} \tag{9}$$

For example, on a gentle slope (Eq. (8)), if $\ell/L = 1/2$, the tsunami run-up height is on the order of $3H$. On a steep slope (Eq. (9)), the run-up height is on the order of H.

The theoretical equation given above does not consider nonlinear effects and energy attenuation by wave-breaking and friction at the ground surface. Considering these factors, the following was proposed as an empirical equation that is applicaple to various types of waves including waves with a large wave height, breaking waves, and solitary waves based on the results of hydraulic model experiments.

$$\log\left(\frac{R}{H}\right) = 0.421 - 0.095\log\left(\frac{\ell}{L}\right) - 0.254\left\{\log\left(\frac{\ell}{L}\right)\right\}^2, (0.1 \le \ell / L \le 1.3) \tag{10}$$

As a tendency, the results of the theoretical equation shown in Eq. (7) are somewhat small at around $\ell/L = 0.1$ and somewhat large at around $\ell/L = 1.0$ in comparison with the results of the empirical equation, Eq. (10). However, these differences are not particularly large. In the experiments that introduced Eq. (10), the maximum run-up height was approximately four times larger than the height of the incident wave.

The following is a run-up solution for a solitary wave on the same topography (Synolakis, 1987).

$$\frac{R}{h} = 2.831(\alpha)^{-1/2}(H / h)^{5/4} \tag{11}$$

However, the wave height H in this equation is the height of the highest water level measured from the mean water level, which is different from the definition of wave height in Eq. (7). The above equation shows good agreement with the experimental values for non-breaking waves, but results in over-evaluation in the case of breaking waves. The run-up height for a breaking wave is $R/h = 0.918(H/h)^{0.606}$ on a slope of approximately 1/20. In an experiment in this connection, the maximum run-up height was also on the order of four times the height of the incident wave.

According to the theory based on which Eq. (7) was introduced, the maximum flow velocity occurs when the wave front crosses the shore line. Its value is

$$u_{\max} = R\sigma/\alpha \tag{12}$$

where, σ is angular frequency and α is the slope of the sea bottom. The distance which a tsunami reaches in the horizontal direction is expressed by R/α.

Now, let us attempt some actual calculations. For example, let us assume a broad continental shelf (horizontal floor) with a water depth of 200 m, and a sloping section with a uniform slope of $\alpha = 1/100$ continuing from this shelf to land. If the period of a tsunami is 10 min and its amplitude is $a = 1$ m, the values obtained from Eq. (8), which approximates a gentle slope, are $R = 7.7$ m and $R/\alpha = 770$ m. In other words, the height of the tsunami is approximately 8 m in the vertical direction, and it reaches a distance of approximately 800 m in the horizontal direction. The maximum flow velocity is $u_{\max} = 8.1$ m/s. This is obviously a quite large tsunami. However, if the uniform slope is $\alpha = 1/10$, the values obtained by applying Eq. (9), which approximates a steep slope, are $R = 2.2$ m and $R/\alpha = 22$ m, and $u_{\max} = 0.23$ m/s. Accordingly, even if an incident tsunami of the same scale approaches from the open sea, the condition of the tsunami when it attacks structures on land will differ greatly depending on the slope that stretches from shallow water to land (or in the larger sense, the topography around the coast). Incidentally, assuming the same cross-sectional topography, and considering the tsunami to be a solitary wave with a wave height of 1 m, the theoretical solution in Eq. (11) gives a run-up height of 7.5 m for a 1/100 slope and 2.4 m for a 1/10 slope. These are not greatly different from the solutions given by Eq. (7).

However, the topography of actual coasts is not a simple uniform slope like that assumed in the foregoing discussion. Therefore, tsunami run-up will also display different behavior. Examples include the following:

(1) The slope on land changes quickly, and the ground height of the inland area is lower than that of coastal sand dunes. In this case, if the tsunami crosses the sand dunes, it will flow toward low-lying areas as a flood flow, and finally will spread over the low ground.
(2) Where the plane topography is complex, the tsunami will concentrate at some location, resulting in a high run-up height in comparison with the surrounding area. Examples of complex topographies include the backs of bays and tips of capes. After a tsunami runs up on land where it is surrounded by mountains and has no place to go, it will collect in valleys. In many cases, this causes an extremely high run-up height in comparison with the surrounding area.
(3) Tsunamis easily invade rivers. It is also important to be aware that a tsunami will run up further in a river than on land due to the fast wave speed of the tsunami in the river.
(4) If the resonance period of a bay coincides with the period of a tsunami, the tsunami will be amplified, resulting in a higher run-up height. In large earthquakes, the fault width is also generally large, and as a result, the period of the tsunami tends to be long. If you look at the records on specific bays, you will find that there have been cases in which the run-up height was larger in a small earthquake when the period of the tsunami coincided with the resonance period of the bay, than in a large earthquake when the period of the tsunami and the resonance period of the bay did not coincide (Hiraishi et al., 1997).

It should also be noted that the locations where run-up height is increased by resonance are not limited to the backs of bays. The locations where run-up height increases vary depending on the surrounding topography, the waveform of the tsunami (in particular, its period), and the wave direction. Thus, it is extremely dangerous to simply believe that “the tsunami will be large in that bay or district, but it's safe here," extrapolating from the very limited number of tsunamis experienced in the past.

References

Shuto, N. (1972): "Standing Waves in Front of a Sloping Dike," Coastal Eng. Japan, Vol. 15, pp. 13-23.

Hiraishi, T., Shibaki, H., and Harasaki, K. (1997): Importance of Resonance Period in Modeling Nankai Earthquake Tsunami, Proc. Coastal Eng., Japan Society of Civil Engineers (JSCE), Vol. 44 (1), pp. 281-285. (in Japanese)

Togashi, H. and Nakamura, T. (1975): Experimental Research on Tsunami Run-up on Land, Proceedings of the 22nd Conference on Coastal Engineering, pp. 371-375. (in Japanese)

Synolakis, C.E. (1987): The Run-up of Solitary Wave, J. Fluid Mech., Vol. 185, pp. 523-545.

(2) *Effects of structures on run-up and inundation*

Normally, in order for a structure to have a protective function against an external force (or a tsunami wave in this case), it must have at least one of the following three functions:

(1) To prevent invasion of the tsunami wave by a structure higher than the height of the tsunami;
(2) To allow partial inflow, but reduce its effect by absorbing it over a large area;
(3) To reduce the energy of the tsunami by friction or whirlpools, supressing its capacity to be transmitted inland.

There are also cases in which one structure has two or more of the above-mentioned functions. For example, a bay-mouth tsunami breakwater is a structure which resists tsunamis using a combination of functions (2) and (3).

However, in the following, the focus is not on large-scale structure specifically constructed as tsunami countermeasures, but on the effect of existing ordinary structures on tsunami run-up and inundation. In this case, ordinary structures include conventional breakwaters, seawalls, and similar structures of comparatively small scale. Actually, for the following reasons, existing ordinary structures cannot be expected to display functions (2) and (3). We will first discuss why it is.

When considering the effect of structures on tsunamis, it is necessary to consider the characteristic features of tsunamis; i.e., their period and wavelength are extremely long. First, let us consider function (3) against tsunamis. Wave-dissipating blocks have effect (3) against ordinary waves, but to secure an adequate wave-dissipating effect, the width of defensive works using wave-dissipating blocks must be a fairly large fraction of the wavelength of the object wave. Thus wave-dissipating works for ordinary waves have a width of several meters. However, a structure of this scale will have virtually no wave-dissipating function against tsunamis, which have a much longer wave length. Effect (3) is conceivable from a breakwater with a narrow opening, which generates a whirlpool at the opening. However, to dissipate the energy of a tsunami, the structure would have to have a scale comparable to a natural cape, like a bay-mouth tsunami breakwater. With a port breakwater of normal scale, only a tiny amount of energy can be dissipated in comparison with the total energy of a tsunami. In other words, unless a structure has a scale like that of a bay-mouth tsunami breakwater, ordinary structures cannot be expected to possess function (3) against tsunamis.

Next, let us consider function (2). Function (2) can be divided into two: a function of maintaining a calm condition in a harbor, and function of allowing shallow inundation over a wide area of land which will suffer little damage if flooded. As actual facilities, breakwaters are designed to provide the former function, and retarding basins the latter function. First, if a breakwater is so large in its scale as a bay-mouth tsunami breakwater, it can be expected to have the former function to some extent. In principle, however, even such a large-scale breakwater is only expected to reduce the volume of inflowing water. However, because the wavelength of a tsunami is longer than the distance to the back of the bay, this type of structure cannot be expected to reduce the effect of a tsunami of the bay over a large area. At ordinary ports and harbors, the area is too small to absorb the impact of a tsunami, and inevitably the port will be

(1) (2) (3) (4)

Fig. 4.13 Kodomari fishing port in the 1983 Nihonkai-Chubu Earthquake (Aomori Prefecture homepage and "Record of 1983 Nihonkai-Chubu Earthquake Disaster").

strongly affected by a tsunami. In fact, in a case shown in the series of photographs in Fig. 4.13, strong currents generated at the opening of breakwaters, affected the port as a whole, and were powerful enough to capsize fishing boats. In short, it is difficult to maintain calm conditions in a harbor with an ordinary breakwater structure. Local weakening of flows is quite possible, but conversely, strong currents will be generated in other locations.

Next, in the latter function in (2), shallow inundation is intentionally allowed over a large area of land which will not be seriously impacted by the inundation. As actual examples, this type of measure was implemented in urban planning after the tsunami disasters at Hilo in Hawaii and at Aonae on Okushiri Island, Japan. At both Hilo and Aonae, the larger part of the tsunami disaster area was converted to parks and the nighttime human population was reduced to zero, basically eliminating the possibility of another disaster. This is the most fundamental type of tsunami countermeasure, and is an idea that should be adopted more widely. In many cases, however, flat coastal land, which is easily inundated by a tsunami, has high utility for everyday life, and as a result, industrial and residential areas are concentrated in such areas. It is difficult to think that cities which have never experienced a tsunami will adopt urban structures which avoid using high-risk areas from the outset of urban planning. Accordingly, it is difficult for existing ordinary structures to have this latter function of (2) against tsunamis.

Following this line of thought, the effect which can be expected from existing ordinary structures is mostly limited to function (1); i.e., to block and prevent invasion of the tsunami, and if the tsunami cannot be blocked completely, reduce the amount of inflowing water.

In this connection, an incident which occurred at Malé Island, the capital of the Republic of Maldives, during the Indian Ocean Tsunami provides a useful reference. Figure 4.14 shows the condition of the tsunami attack around Malé Island, as reproduced by numerical calculation. The position of Malé Island is indicated by an arrow in Fig. 4.14. Malé Island is a small island in the North Malé Atoll. The tsunami attacked from the east. Because the atoll itself acted as a barrier to the tsunami, the tsunami concentrated in the Vaadhoo Channel between the North Malé Atoll and South Malé Atoll. It is thought that the water level of the tsunami increased somewhat on the eastern coast of Malé Island, which was hit directly by the tsunami, and on the southern coast facing the Vaadhoo Channel. According to a wave gauge located in the atoll at Malé International Airport, the maximum water level during the 1st wave of the tsunami attack was 2.06 m above the datum sea level, and the maximum variation attributable to the tsunami was 1.45 m. Because the datum level is 0.64 m lower than mean sea level, the tide level (sea level) at the time of the tsunami attack was roughly the same as mean sea level. Figure 4.15 shows the crown heights of the seawalls and detached breakwaters and the land height at Malé Island, and the inundated area in the tsunami. The crown height of

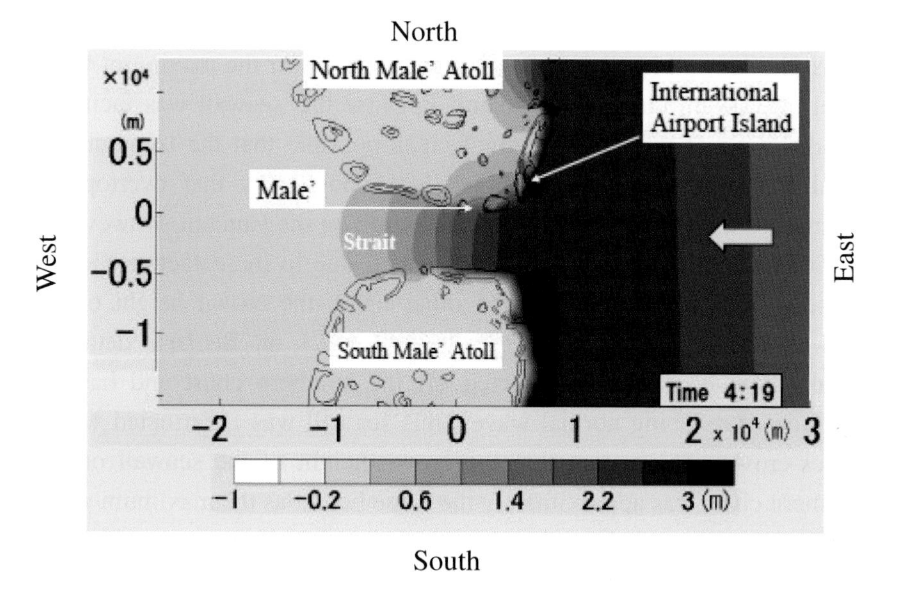

Fig. 4.14 Tsunami attack at Malé Island (Ohtani et al., 2005).

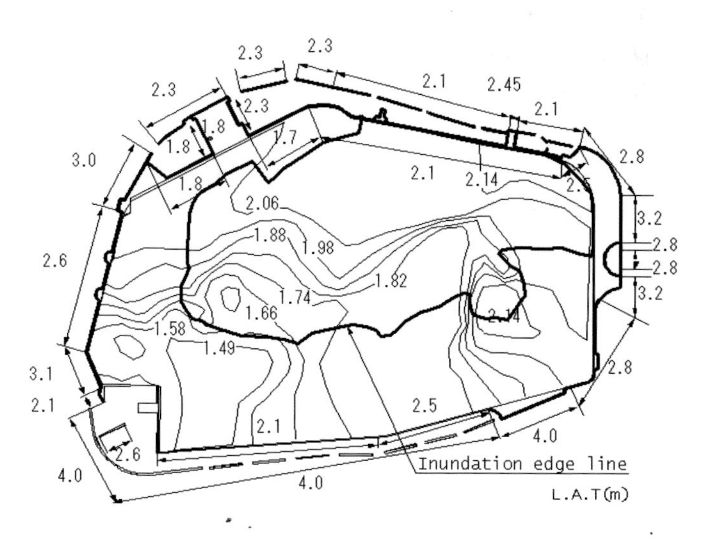

Fig. 4.15 Crown heights of seawall and detached breakwater, land height, and inundation area at Malé (Fujima et al., 2005).

the seawall on the southeastern coast is 2.8 m above datum sea level. Thus, the seawall has a margin of about 70 cm over the maximum water level of 2.06 m during the tsunami. Because this seawall was facing in the direction of the tsunami attack, it is possible that the tsunami was amplified somewhat, and there is also a possibility that overtopping occurred due to swells or wind waves riding on the tsunami. However, it is thought that overtopping of this seawall due to these factors did not cause serious inundation. On the other hand, the crown height of the seawall on the southern coast is lower, at 2.1 m. Because detached breakwaters had been constructed off the southern coast and has the function of reducing normal waves, this seawall was constructed with a lower crown. This meant that the crown height of the seawall on the southern coast was approximately the same height as the maximum water level of the tsunami. It is also estimated that the tsunami was amplified somewhat at the southern coast. Therefore, it is thought that the main body of the tsunami overtopped the seawall, causing inundation. The height of the quay wall (anchorage) in the port on the northwestern side of the island is 1.8 m. It is thought that this structure was also overtopped, and inundation spread from this point as well.

Figure 4.16 shows the results of numerical computations of the effects of these various structures. (a) Case 1 is the case calculated under the actual conditions at the time of the tsunami attack; that is, considering both the detached breakwaters and seawalls. The inundated area shows good agreement with the actual measured values. (b) Case 2 is the case without the existing detached breakwaters on the southern coast. Although the inundated area increases slightly, the results are substantially the same as in Case 1. In (c) Case 3, which assumes that neither seawalls nor the detached breakwaters exist, the inundated area expands greatly. Figure 4.17 shows the distribution of flow velocities in Case 1 and Case 3. Focusing not on the flow velocity vectors in the figures, but on the color coding showing flow velocity, it is apparent that, at the position of the seawall, there are some areas where the flow velocity is faster with the seawall than without it, but at virtually all locations on land, the flow velocity is reduced in the case with the seawall. In other words, the "damming" effect of the seawall on the southern coast reduced both the velocity and the amount of inundation, and thus was useful in damage mitigation.

Returning once again to Fig. 4.16, the following explains the reason for examining (d) Case 4. The tsunami attacked during a time period when the tide level (sea level) was substantially at the same level as the mean sea level. However, high tide occurred 14 hours later, and the sea level rose by approximately +0.5 m. Therefore, (d) Case 4 shows the

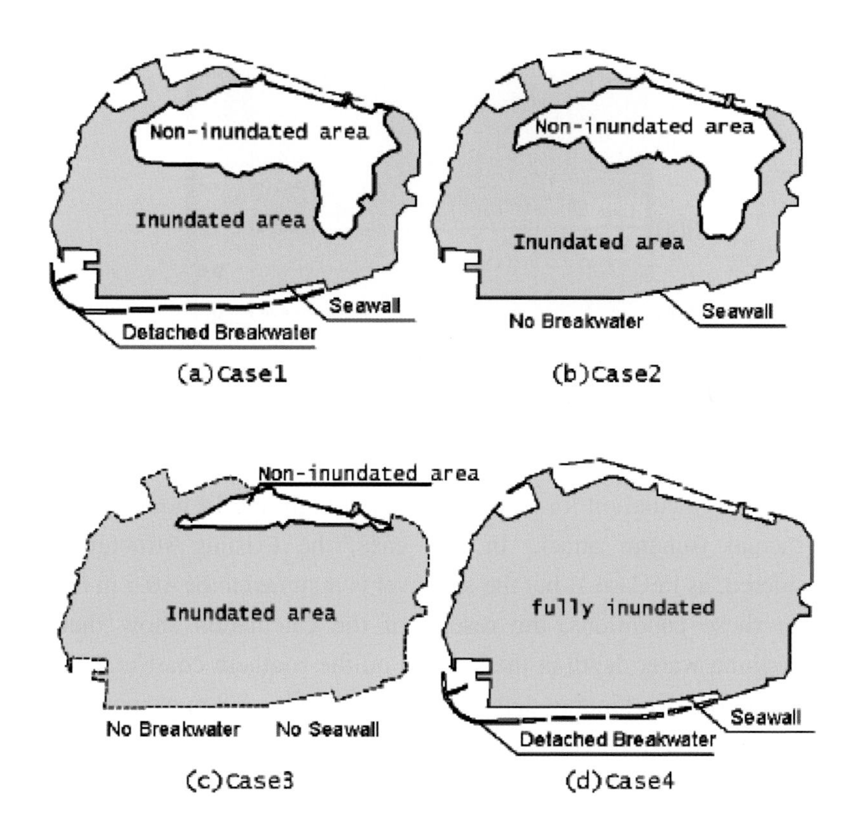

Fig. 4.16 Results of inundation computations for Malé Island ((a) Case 1: Considering existing seawalls and detached breakwaters, (b) Case 2: Ignoring the detached breakwaters on the southern coast, (c) Case 3: Ignoring both the seawalls and detached breakwaters, (d) Case 4: Considering the existing structures, but assuming the sea level was mean sea level + 0.5 m) (Ohtani et al., 2005).

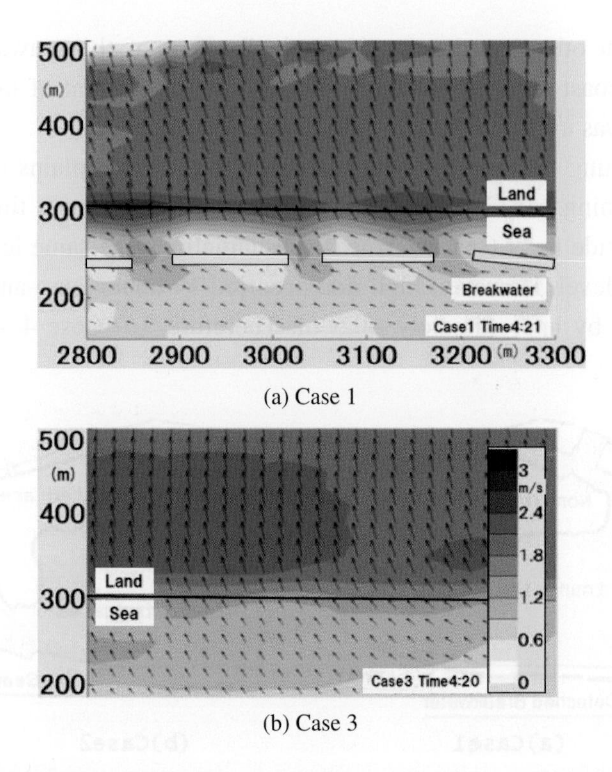

Fig. 4.17 Distribution of flow velocity in (a) Case 1 and (b) Case 3 (Ohtani et al., 2005).

results of a calculation for tsunami attack at high tide 14 hours later than the actual tsunami attack. In this case, the existing structures are considered, as in Case 1, but the sea level is assumed to be +0.5 m higher. Under these conditions, the results of the calculation show that the overtopping water depth at the seawall on the southern coast is 0.5 m or more, and the overtopping depth at the quay wall is 0.8 m or more. Thus, the tsunami overtops coastal structures around much of the island, and the entire islands suffer inundation damage. If the height of a tsunami greatly exceeds the crown height of structures, the final overtopping water depth will be large, and as a result, seawalls will be ineffective in reducing the tsunami run-up height and limiting inundated area. At least, however, they have an effect of delaying the inundation until the water level exceeds the crown height, and this time can be used for evacuation.

Summarizing this discussion, in order for existing ordinary structures (structures not designed for tsunami disaster prevention) to demonstrate a function of reducing the run-up height of a tsunami and limiting the inundated area by a blocking or "damming" effect (function (1) mentioned at the beginning of this section), it is a precondition that the height of the tsunami must not exceed the height of the structure, or must exceed the height of the structure by only a very small amount. In others words, provided a tsunami is not of very large scale, a disaster prevention function against tsunamis can also be expected from ordinary seawalls and similar coastal structures. However, it is necessary to consider the fact that ordinary structures will be of little or no use in limiting the inundated area or reducing the run-up height in a tsunami attack in which the height of the tsunami greatly exceeds that of the structure.

In order to be effective against large tsunamis, structures must also be large. The Hiromura Bank in Hirogawa Town, Wakayama Prefecture, Japan is proof that the inundated area can be limited if an embankment which exceeds the height of a tsunami is constructed (see Chapter 2.5).

References

Fujima, K., Tomita, T., Honda, K., Shigihara, Y., Nobuoka, H., Hanazawa, M., Fujii, H., Ohtani, H., Orishimo, S., Tatsumi, M., and Koshimura, S. (2005): "Preliminary Report on the Survey Results of the 26/12/2004 Indian Ocean Tsunami in the Maldives," p. 89.

Ohtani, H., Fujima, K., Shigihara, Y., Tomita, T., Honda, K., Nobuoka, H., Koshimura, S., Orishimo, S., Tatsumi, M., Hanzawa, M., and Fujii, H. (2005): The Inundation Characteristics of Malé Island and Airport Island and the Effects of Seawalls and Detached Breakwaters in the Maldives due to the Indian Ocean Tsunami, Annual Journal of Coastal Engineering, Japan Society of Civil Engineers (JSCE), Vol. 52, pp. 1376-1380. (in Japanese)

Japan Meteorological Agency:
http://www.seisvol.kishou.go.jp/eq/inamura/p1/html

Aomori Prefecture:
http://www.bousai.pref.aomori.jp/jisinsouran/nihonkai/select_menu.htm

(3) *Tsunami wave force*

The size of tsunami wave force varies greatly. The size of the tsunami wave force acting on structures varies depending on various conditions, including not only the size of the tsunami but also the tsunami waveform, the seafloor topography, and other factors, and also changes over the course of time. Figure 4.18 is a schematic illustration of the change over time in tsunami wave force. Tsunami wave force can be divided into the wave force of the bore and the wave force of the standing wave. The wave force of a bore is a dynamic load that occurs when the front of a tsunami impacts on a structure. Its peak value is defined as the impact load of the bore. The value of the impact load of the bore changes greatly depending on conditions, even assuming the same tsunami height. When a bore-shaped tsunami acts in comparatively shallow water or on land, in some cases the value of the impact load of the bore is extremely large. On the other hand, when the height of the tsunami is small or the tsunami acts without breaking in comparatively deep water, the wave force of the bore does not become particularly large in comparison with the wave force of the standing wave, and in some cases no remarkable peak can be observed. The wave force of the standing wave is a quasi-static load caused by the increase in water depth, and is fundamentally proportional to the water depth at the front side of a structure.

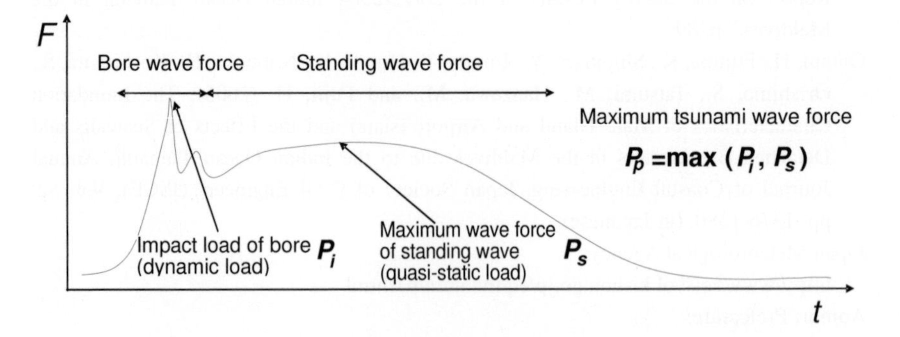

Fig. 4.18 Schematic diagram of change over time in tsunami wave force (Arikawa et al., 2005).

The following will present a brief outline of tsunami wave forces, discussing the tsunami wave force acting on structures in the water, such as breakwaters and seawalls, and the tsunami wave force acting on structures on land, such as houses and other buildings. Several equations for tsunami wave force are introduced below; in all cases, these have been proposed for use in the design of structures. However, caution is required, as the value of the wave force of a tsunami not only varies greatly, depending on various conditions, but also shows spatial and temporal deviations. As a result, the wave force obtained using these equations may not necessarily give an accurate evaluation of the actual tsunami wave force.

(1) *Wave force of tsunamis acting on breakwaters and seawalls*

The tsunami wave force acting on vertical walls in the sea, such as breakwaters and seawalls, can generally be obtained using the wave pressure distribution as shown by the following equation (Tanimoto et al., 1983). As shown in Fig. 4.19, the wave pressure above the still water level displays a linear distribution with $p = 0$ at a height of $\eta^* = 3.0a_I$ becoming $p = 2.2\rho_0 g a_I$ at still water level; below the still water level, the pressure assumes a uniform distribution of $p = 2.2\rho_0 g a_I$.

$$\eta^* = 3.0a_I \quad (13)$$

$$p_1 = 2.2\rho_0 g a_I \quad (14)$$

$$p_u = p_1 \quad (15)$$

where,

η^* : Height of wave pressure acting above still water level (m)

a_I : Height of incident tsunami above still water level (amplitude of tsunami) (m)

p_0 : Density of sea water (t/m^3)

g : Acceleration of gravity (m/s^2)

p_1 : Intensity of wave pressure at still water level (kN/m^2)

p_u : Uplift pressure at bottom edge of front side (kN/m^2)

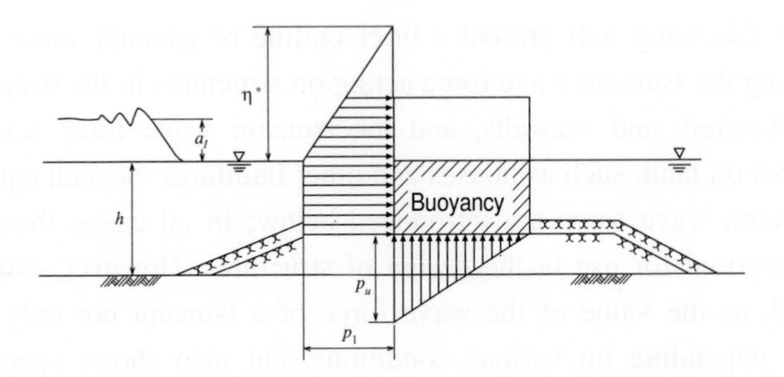

Fig. 4.19 Distribution of tsunami wave force acting on a structure in the sea.

The still water level in case of a bore-type tsunami wave is the water level immediately before the attack by the bore. In non-breaking tsunami waves, the height H_I of the incident wave of the tsunami is,

$$H_I = 2a_I \tag{16}$$

In this equation, the equation for wave force when a normal wave acts on a vertical wall is applied to tsunamis, which have a long period. The tsunami wave force obtained using this equation corresponds to the maximum wave force of a standing wave shown in Fig. 4.18.

On the other hand, in cases of shallow coasts where an extremely gentle seafloor slope continues over a considerable distance, as described in 2.1, soliton fission can easily occur as a result of nonlinearity and dispersion effects. Because the wave height increases in this kind of tsunami accompanied by soliton fission, in many cases, the wave force is larger than when a tsunami does not undergo soliton fission. Based on experimental results, the following equation for tsunami wave force has been proposed for tsunamis accompanied by soliton fission (Ikeno et al., 1998). Namely, above the still water level, $p = 0$ at a height of $\eta^* = 6.0a_I$; the wave pressure then displays a linear distribution, becoming $p = 3.5\rho g a_I$ at height $\eta^* = 3.0a_I$ and below $\eta^* = 3.0a_I$, assumes a uniform distribution of $p = 3.5\rho g a_I$.

$$\eta^* = 6.0a_I \quad (17)$$

$$p_1' = 3.5\rho g a_I \quad (18)$$

$$p_u = p_1 \quad (19)$$

where,

p_1' : Intensity of wave pressure at $\eta^* = 3.0a_I$ above still water level (kN/m^2).

It may be noted that the tsunami wave force obtained with this equation is close to the upper limit value of the impact load of a bore in Fig. 4.18.

(2) *Wave force of tsunamis acting on structures on land*

In cases where a tsunami runs up on land, the tsunami wave force is generally expressed using the run-up water depth in a condition in which no structures exist, η_{max}, as an index. As the horizontal wave force of a tsunami acting on a structure on land, the wave pressure distribution shown in Fig. 4.20 has been proposed for cases in which soliton fission does not occur. The value of the coefficient α varies depending on the run-up flow velocity u_{max}. Assuming $\alpha = 3.0$, the result is substantially the upper limit of the maximum horizontal wave force of a tsunami in case of a wave without soliton fission. When the Froude number ($F_r = u_{max}/\sqrt{g\eta_{max}}$) in tsunami run-up becomes closer to 1.0, the value

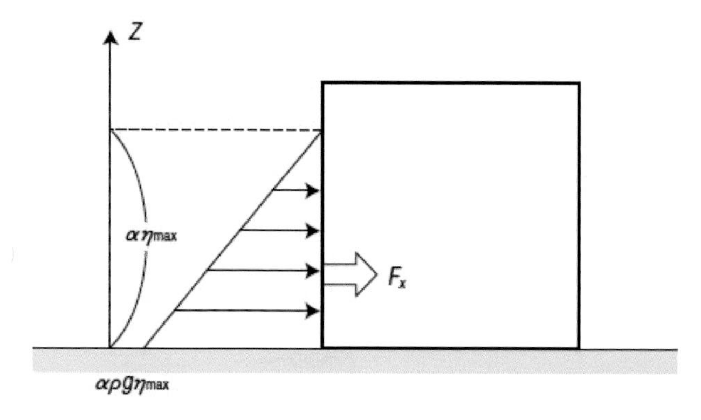

Fig. 4.20 Distribution of tsunami wave force acting on a structure on land (wave without soliton fission) (Asakura et al., 2000).

of α will also becomes closer to 1.0 (static water pressure corresponding to run-up depth).

These wave forces are considered to be equivalent to the maximum wave force of a standing wave in Fig. 4.18. According to the results of another experiment, an extremely large bore impact load may act on structures, depending on the conditions. In some cases, the local wave pressure may reach more than five times the static water pressure corresponding to run-up depth (Arikawa et al., 2006). However, various factors affects the results. For example, this impact load changes greatly depending on the strength (rigidity) of the structure. In addition, the time of action is extremely short, at 0.5 s or less when converted to real site condition. Therefore, the extent to which this contributes to the actual destruction of structures is unclear at present.

On the other hand, when a tsunami undergoes soliton fission, the wave force of the tsunami increases in comparison with a wave that does not under fission, in the same way as at sea. In cases where soliton fission occurs, a particularly large wave force acts at the bottom of structures. Considering this tendency, a wave force distribution in which an increment of wave pressure due to soliton fission is added to the wave distribution of a non-soliton wave has been proposed as shown Fig. 4.21.

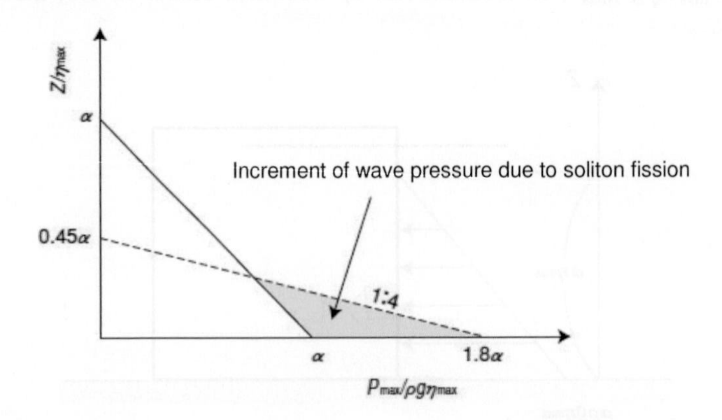

Fig. 4.21 Distribution of tsunami wave force acting on a structure on land (wave with soliton fission) (Asakura et al., 2000).

References

Arikawa, T., Ikebe, M., Yamada, F., Shimosako, K., and Imamura, F. (2005): Large Model Test of Tsunami Force on a Revetment and on a Land Structure, Annual Journal of Coastal Engineering, JSCE, Vol. 52, pp. 746-750. (in Japanese)

Tanimoto, K., Takayama, T., Murakami, K., Murata, S., Tsuraya, H., Takahashi, S., Morikawa, M., Yoshimoto, Y., Nakano, S., and Hiraishi, T. (1983): Actual Condition of the 1983 Nihonkai-Chubu Earthquake Tsunami and Second and Third Considerations, Technical Note of the Port and Harbour Research Institute, No. 470, pp. 299.

Ikeno, M., Matsuyama, M., and Tanaka, H. (1998): Experimental Research on Deformation of Tsunamis with Soliton Fission on Continental Shelf and Design Wave Pressure of Breakwaters for such Tsunamis, Proceedings of Coastal Engineering, JSCE, Vol. 45, pp. 366-370. (in Japanese)

Asakura, R., Iwase, K., Ikeya, T., Takao, M., Kaneko, T., Fujii, N., and Omori, M (2000): An Experimental Study on Wave Force Acting on On-Shore Structures due to Overflowing Tsunamis, Proceedings of Coastal Engineering, JSCE, Vol. 47, pp. 911-915. (in Japanese)

Arikawa, T., Ohtubo, D., Nakano, F., Shimosako, K., Takahashi, S., Imamura, F., and Matsutomi, H. (2006): Large Model Test on Surge Front Tsunami Force, Annual Journal of Coastal Engineering, JSCE, Vol. 53, pp. 796-800.

Chapter 5

Tsunami Simulations and Forecasting Systems

5.1 Tsunami Simulations

(1) *Predicting tsunami behavior*

It would be needless to say that quicker and more accurate tsunami forecasting is important for mitigation the disaster, in particular, loss of life and injury. Because the behavior of tsunamis changes in a complex manner due to effect of coastal topography, numerical models obtained analytically on the basis of simple assumptions are inadequate. In order to know the actual behavior of a tsunami and estimate the height of the tsunami, it is necessary to predict the propagation of the tsunami in a way that fully incorporates the seafloor topography. The only method for achieving this is a simulation using the actual initial waveform of the tsunami and topographical data. Research activities which are conscious of tsunami forecasting have been carried out since around 1983. And the possibility of practical tsunami simulations was first pointed out after seafloor topographical data was started to be collected off Japan's Sanriku coast. One work has shown that the maximum water level distribution can be reproduced with accuracy within an error of 20% when the rupture process (fault model) in past earthquakes is relatively well understood (Shuto et al., 1988).

Because the basic equation for tsunami simulations is a time-dependent model, when the initial conditions are given, it is possible to obtain a variety of information on the behavior of a tsunami, including its arrival time at object locations, height of the 1st wave, maximum wave height, duration, velocity field, etc. However, in time-dependent models (particularly in explicit differential schemes), it is necessary to satisfy stability conditions, and if these conditions are not satisfied, the simulation will give unrealistic values. Depending on the selection of the computational grid (mesh) and the differential scheme, discretization error may occur in some cases, resulting in reduced predictive accuracy (Imamura et al., 1986; Imamura and Goto, 1988). However, if the time required for the simulation is longer than the time from the generation of a tsunami to its arrival at the coast, this method is not useful in real time tsunami prediction where computations begin simultaneously with the occurrence of an earthquake. In such cases, shortening the computation time becomes an important study item, on the same level as accuracy. Another approach to tsunami prediction is to perform computations for a large number of cases giving initial values in advance, and accumulate the results in a database for future utilization. Because simulations are performed before an earthquake occurs, this approach relieves the constraint of computational speed.

On the other hand, the earthquake and tsunami observation network has gradually been expanded, making it possible to carry out detailed investigations of the actual condition of tsunamis. As a result, it is now known that the tsunami wave source estimated from a fault model by seismic wave analysis does not necessarily coincide with the wave source estimated from tsunami observation or investigation. Recently, the importance of dynamic fault motion behavior, which had been ignored until now in tsunami simulations, has also been pointed out. Furthermore, precise estimation of the tsunami source area in real time is also a difficult task. Study of a real-time method of tsunami prediction which integrates simulation and the tsunami observation network was begun in order to solve these wide-ranging problems. Provided that the attacking tsunami can actually be captured directly by offshore observation, it should be possible to use that data as input information for a reliable tsunami simulation, even when the fault motion and wave source have some unknown factors, and further even when it is a tsunami caused by phenomena other than earthquakes. This requires numerical forecast is performed simultaneously with real-time observation using an advanced tsunami monitoring system, where much more improvement is made with regard to the location of tsunami meters (GPS buoys), their number, and so forth.

(2) *Dissemination of simulation techniques*

The activities to provide simulation techniques developed in Japan to other tsunami-prone countries would bring about significant benefit to these countries. As efforts during the International Decade for Natural Disaster Reduction (IDNDR; 1987-1997), the TIME (Tsunami Inundation Modeling Exchange) Project was launched with the support of a joint project of the International Union of Geodesy and Geophysics (IUGG) and UNESCO's Intergovernmental Oceanographic Commission (IOC). In the TIME Project, simulation techniques were transferred to countries which have suffered tsunami damage in the past, or are expected to suffer such damage in the future (Fig. 5.1). The source code was made available, and a complete manual was prepared and distributed (IOC, 1997). In addition, local researchers acquired Tohoku University's tsunami numerical model code (TUNAMI: Tohoku University Numerical Analysis Modeling for Inundation), and are inputting topographical and land use data for areas of concern, data on assumed earthquakes, and other information, calculating expected tsunami inundation, and preparing

Fig. 5.1 Logo of the TIME Project.

hazard maps, etc., while also conducting training and joint research. As of 2007, 39 organizations from 21 countries have participated in this project. Resolutions expressing appreciation for this project have repeatedly been passed at various conferences, including those of the IUGG and groups involved in actual tsunami warning work.

References

Imamura, F. and Goto, C. (1986): Truncation error in numerical simulation by the finite difference method, Journals of the Japan Society of Civil Engineers, No. 375/II-6, pp. 241-250. (in Japanese)

Imamura, F. and Goto, C. (1988): Truncation error in numerical simulation by the finite difference method, Coastal Engineering in Japan, Vol. 31, No. 2, pp. 245-263.

Shuto, N., Goto, C., and Imamura, F. (1988): Use of numerical simulation in tsunami forecasting and warning, Journals of the Japan Society of Civil Engineers, No. 393/II-9, pp. 181-189. (in Japanese)

IOC (1997): Numerical method of tsunami simulation with the leap-frog scheme, IUGG/IOC TIME Project, Manual and Guides 35, SC-97/WS-37.

5.2 Tsunami Forecasting Systems and Their Evolution

(1) *Features of tsunami forecasting*

Tsunamis can be classified as "local" or "remote," depending on the distance from the source of the tsunami to the coast concerned. This classification is based on relative distance and is not a classification based on characteristics of tsunamis themselves. For example, when the tsunami caused by the sea bottom earthquake off the coast of Chile in South America arrived at the coast of Chile, it was considered a "local tsunami," but when the same tsunami propagated across the Pacific Ocean, reaching Japan on the opposite side, it was a "remote tsunami." Because local tsunamis occur close to the focus of the earthquake, the ground motion caused by the earthquake can frequently be felt by residents along the coast, whereas in remote tsunamis, it is impossible to feel the ground motion due to the distance from the center of the earthquake.

In local tsunamis, rapid evacuation is indispensable because the time from occurrence to arrival of the tsunami is short. Therefore, in forecasts for local tsunamis, the response from the occurrence of the earthquake to issuance and transmission of the tsunami forecast must be rapid. The key point is how to minimize the time required for this.

On the other hand, in the case of remote tsunamis, there is a considerable time margin between the occurrence of the tsunami and its arrival. Therefore, in forecasts for remote tsunamis, the highest priority is not to shorten the time required to issue the forecast. Rather, because remote tsunamis propagate over an extremely large area, it is critically important to strengthen the earthquake observation system and create a network that ensures proper transmission of warnings by cooperation between coastal nations, corresponding to the process of occurrence and propagation of tsunamis.

Because the issues related to tsunami forecasting differ in local and remote tsunamis, as outlined above, this section will discuss the evolution of tsunami forecasting systems separately for the two types of tsunamis. In the case of local tsunami forecasting, efforts in Japan will be described. For remote tsunami forecasting, the history of efforts in the Pacific Ocean region and the current status of those efforts, occasioned by the 2004 Indian Ocean Tsunami, will be described.

(2) *Local tsunami forecasting*

An organization responsible for tsunami forecasting was created in Japan in 1941 for the Sanriku coast of northeastern Japan, which had suffered numerous tsunami attacks in the past. However, tsunami forecasting on a nationwide scale based on the official legal framework began only in 1952 in the Central Meteorological Observatory (now the Japan Meteorological Agency). In the initial stage of forecasting work, the arrival times of earthquake P waves and S waves observed at various earthquake monitoring points were transmitted by telephone or telegram to the central tsunami forecasting office, where the focus and magnitude of the earthquake were determined manually using maps, charts, pencils, compasses, rulers, and certain types of calculators. The person in charge then judged whether or not a tsunami would occur by referring to a "Tsunami forecasting diagram" prepared empirically based on past data in such a way that the possibility of a tsunami could be judged from the relationship among the intensity of the earthquake, duration of the preliminary tremor, and total amplitude. On average, the time until forecasts were issued was around 17 minutes. Further, only means available to communicate the resultant judgment to the region where the tsunami attack was expected were telegram and telephone.

To shorten the time required to issue tsunami forecasts, first, the Japan Meteorological Agency (JMA) improved the communications network for collecting earthquake observation data, automated data processing, etc. in order to speed up the process of determining the focus and magnitude of earthquakes as far as possible. As a result, the time required for tsunami forecasts was reduced to approximately 14 minutes in 1980. Three years later, when the Nihonkai-Chubu Earthquake occurred on May 26, 1983, the JMA succeeded in issuing a tsunami warning 14 minutes after the earthquake, as planned. However, the tsunami arrived at the coast even before the forecast was issued. The 1st wave attacked only seven minutes after the earthquake, and 100 people died in the tsunami.

Therefore, with the aim of issuing forecasts within seven minutes after an earthquake occurs, the JMA introduced a system that automatically detects earthquakes and automatically performs the entire series of processing operations from calculation of the location and magnitude of the earthquake to judgment of the possibility of a tsunami and issuance of the judgment. At the same time, tsunami information is also issued by emergency broadcasting by news organizations. This system actually operated during the Hokkaido Nansei-oki Earthquake on July 12, 1993, and issued a tsunami warning in only five minutes, which was faster than the target time. Unfortunately, however, a tsunami with a maximum height of 10 m struck Okushiri Island off the Hokkaido coast only three minutes after the earthquake, killing 198 persons.

Following this disaster, the existing monitoring network was reviewed in April 1994 targeting a further shortening of the time required to judge whether a tsunami will occur or not. The tsunami earthquake early detection network was expanded by installing seismographs at approximately 180 locations throughout Japan, and the time required to issue tsunami forecasts was reduced to three minutes. The current tsunami forecasting system began operation in April 1999 using quantitative tsunami forecasts, and incorporates achievements in tsunami propagation simulation techniques in recent years. Approximately 100,000 simulations of the condition of tsunami generation and propagation after earthquakes were performed in advance using various earthquake locations, depths, and magnitudes, and the results were accumulated in a database. When an earthquake occurs, the system searches these simulation results for earthquakes which resemble the actual earthquake, and announces the height and other behavior of the tsunami expected in various areas. In comparison with the conventional technique of prediction from past examples based on empirical rules, this system uses the results of actual model calculations that cover the entire process from the occurrence of a tsunami caused by fault motion in an earthquake to the transmission of the tsunami wave to the coastline. This has made it possible to divide the forecasting areas into much smaller segments and describe the height of predicted tsunamis more in detail.

The evolution of the tsunami forecasting system in Japan up to the present has been described here. The present flow of preparation and transmission of earthquake tsunami information is shown in Fig. 5.2.

Beginning on August 1, 2006, the Japan Meteorological Agency began providing "Earthquake Early Warnings (EEW)," which precede the arrival of the main earthquake motion. This "lead time to strong motion" is expected to be useful information for disaster prevention. EEW are prepared by rapidly analyzing the observation data obtained by seismographs located near the focus of an earthquake and estimating the arrival time/seismic intensity in various areas before the main earthquake motion arrives. In some earthquake, it has become possible to shorten the time to issuance of a tsunami forecast to less than two minutes at the fastest

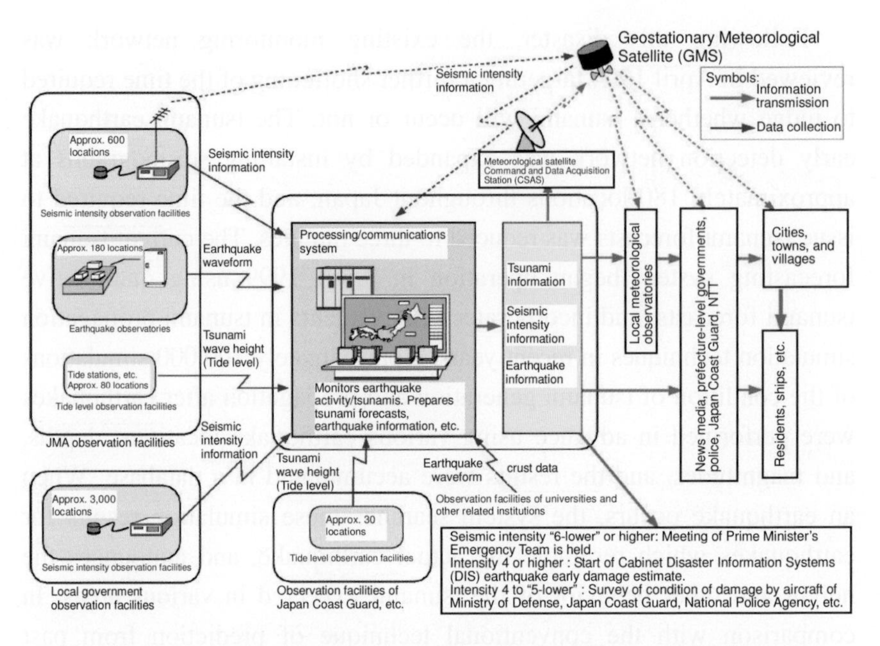

Fig. 5.2 Flow of preparation and transmission of earthquake tsunami information.

by using the automatic processing technology of this EEW system (see Fig. 5.3). The tsunami forecasting function has been operational since October 2006. As this example shows, the Japan Meteorological Agency is continuing its efforts to achieve faster tsunami forecasting. In the Noto Peninsula Earthquake (*M*6.9) on March 25, 2007, a tsunami forecast was issued 1 min 40 s after the earthquake.

Recent examples have revealed new challenges for quantitative tsunami forecasting based on a database containing the results of approximately 100,000 simulations. As one such example, during an earthquake (*M*7.9) east of the Kuril Islands, which occurred just after 8:00 p.m. on November 15, 2006, tsunami warnings/alerts were issued for the coast of Japan based on this database. All of these tsunami forecasts were lifted by 1:30 a.m. on November 16. However, the largest wave height of this tsunami was recorded at Miyake Island at 4:09 a.m. in the same morning. In other words, the maximum tsunami height was observed after the tsunami forecast had been cancelled. It is estimated

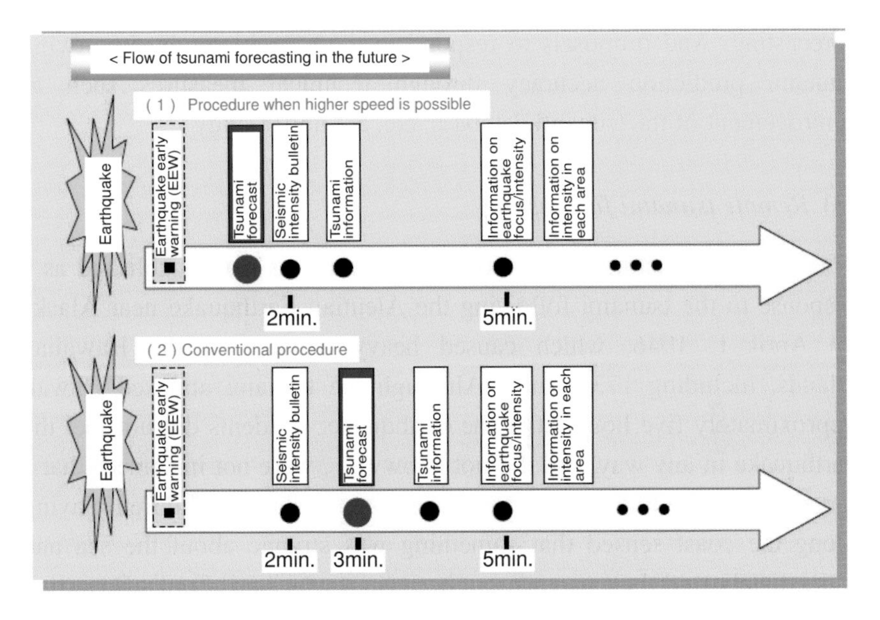

Fig. 5.3 Flow of tsunami forecasting in the future.

that this largest wave was a reflected wave (scattered wave) from the Emperor Seamount Chain, which is a sea bottom mountain range to the east of the focus of the earthquake, and this had not been considered in simulations. On January 13, 2007, another earthquake occurred at basically the same location and focal depth as the earlier Kuril earthquake. Because the magnitude of the 2007 earthquake ($M8.2$) was larger than that of the 2006 earthquake, the tsunami height in tsunami warnings/alerts was also larger than in the previous tsunami. However, the actual tsunami height was smaller. When the fault used in the simulation was compared with the actual fault, it was found that the simulation conditions were set to cause a larger tsunami. In general, the tsunami database is based on the predicted values for the worst case assumed from the location of the center and the magnitude of the earthquake.

These examples highlight the need to ensure that all persons concerned properly understand that the predicted values contain some degree of error, corresponding to these problems in quantitative tsunami forecasting. And proposals to respond to these problems by improving tsunami prediction accuracy through technical measures such as improvement of the tsunami database, etc. are under study.

(3) *Remote tsunami forecasting*

A tsunami forecasting system for remote tsunamis was established as a response to the tsunami following the Aleutian Earthquake near Alaska on April 1, 1946, which caused heavy damage in the Hawaiian Islands, including 173 deaths. Although the tsunami attacked Hawaii approximately five hours after the earthquake, residents did not feel the earthquake in any way, and did not know (i.e., were not informed) that a large earthquake had occurred in the Aleutian Islands. Children playing along the coast sensed that something was strange about the sea and immediately told their parents, but because it was April Fool's Day, their parents did not heed the warning. It is said that this contributed to the heavy loss of life.

As a result of this experience, the United States established the Pacific Tsunami Warning Center (PTWC) in 1949 as a tsunami warning organization, centering on the Coast and Geodetic Survey (CGS). However, because this was a domestic system in the US, there was no cooperation between the various countries around the Pacific Ocean during the Chilean Earthquake Tsunami in 1960. As a result, many of the nations around the Pacific suffered heavy damage, beginning with the Hawaiian Islands and also including Japan. Based on this, in 1965, the International Tsunami Information Center (ITIC) was established under UNESCO's Intergovernmental Oceanographic Commission (IOC) as an organization which provides information on tsunami warning systems and supports the construction of systems and tsunami disaster mitigation planning in each nation, with the aim of alleviating damage due to tsunamis in the Pacific coastal nations. In 1968, the International Coordination Group for the Tsunami Warning System in the Pacific (ICG/ITSU) was established under the same UNESCO IOC as a warning organization for remote tsunamis occurring in the Pacific Ocean.

(The name of this organization has now been changed to the Intergovernmental Coordination Group for the Pacific Tsunami Warning System (ICG/PTWS).) This organization was established to strengthen the tsunami disaster prevention systems of the Pacific nations by exchanges and sharing of information on earthquakes and tsunamis occurring in the Pacific Ocean among the member nations. The group now has 30 member nations, including Japan. The ICG/ITSU holds meetings every two years in order to coordinate efforts among the member nations and give recommendations on tsunami warning systems. The previously-mentioned Pacific Tsunami Warning Center in Hawaii is conducting the operation for actually providing tsunami information to the member nations.

Following this, a proposal by the Japan Meteorological Agency (JMA) to establish a tsunami warning center for the northwestern Pacific region was approved at the 1999 meeting of the ICG/ITSU. Since March 2005, the JMA has provided "northwestern Pacific tsunami information" to the various related countries around the northwestern Pacific Ocean as the Northwest Pacific Tsunami Advisory Center (NWPTAC). At present, NWPTAC provides information to 10 nations. This "northwestern Pacific tsunami information" is issued when an earthquake of magnitude 6.5 or greater occurs in the northwestern Pacific region, and comprises the time when the earthquake occurred, the location of the earthquake focus, the magnitude of the earthquake, the estimated possibility of a tsunami, and when there is a possibility of a tsunami, the estimated arrival time and predicted height of the tsunami at points on the coast. The "northwestern Pacific tsunami information" issued by the JMA is used by the recipient nations to take emergency tsunami disaster prevention action, such as issuing domestic tsunami warnings for the predicted tsunami in the country concerned, advising residents to evacuate, and similar actions. In this connection, it may be noted that the tsunami information issued by the Pacific Tsunami Warning Center does not include the predicted height of tsunamis.

The Great Indian Ocean Tsunami, which was caused by a giant earthquake off the northwest coast of Sumatra Island in Indonesia on

December 26, 2004, was an unprecedented disaster, claiming more than 300,000 lives in coastal countries across the entire Indian Ocean region. As one factor in the large-scale damage, it has been pointed out that no tsunami warning system like that in the Pacific Ocean had been created for the Indian Ocean. Utilizing its experience in issuing tsunami information in the Pacific region, the Japan Meteorological Agency, in cooperation with the Pacific Tsunami Warning Center, began providing tsunami monitoring information in March 2005 to 26 nations around that Indian Ocean which requested this service. This is being carried out as a provisional action until an actual Indian Ocean tsunami warning system is constructed. "Indian Ocean tsunami information" is issued when an earthquake of magnitude 6.5 or greater occurs in the Indian Ocean region, and comprises the time, location of the focus, and magnitude of the earthquake, the possibility of a tsunami, and when a tsunami is possible, the predicted value of the time until arrival of the tsunami at designated coastlines.

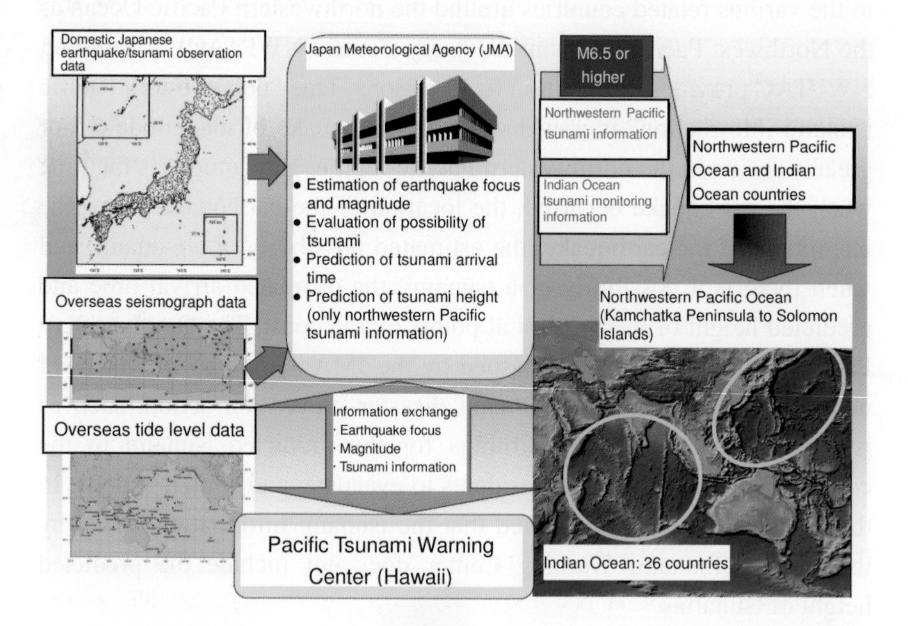

Fig. 5.4 Flow of northwestern Pacific tsunami information and Indian Ocean tsunami monitoring information.

The flow of tsunami information provision in the northwestern Pacific Ocean and Indian Ocean under the responsibility of the Japan Meteorological Agency is shown in Fig. 5.4.

In June 2005, the Intergovernmental Coordination Group for the Indian Ocean Tsunami Warning and Mitigation System (ICG/IOTWS) was established with the participation of the Indian Ocean nations, the members of UNESCO's Intergovernmental Oceanographic Commission, and others. This group is studying an implementation plan for the construction of an Indian Ocean tsunami warning system. The actual study is being carried out by six working groups which were set up for individual topics such as earthquake monitoring, tide level monitoring, disaster risk evaluation. Early progress in strengthening the tsunami warning systems in the respective countries around the Indian Ocean is expected.

5.3 Future of Tsunami Forecasting Systems

Although the importance of information in disaster prevention is steadily increasing, the importance of information in connection with tsunamis is particularly large. Although the time margin between an earthquake and a tsunami attack is small, lives can be saved by evacuation during this interval. However, not all earthquakes which are accompanied by tremors cause tsunamis, and there is a type of earthquake that causes large tsunamis even though the tremors are small. Thus, it is difficult for residents to judge that evacuation is necessary based simply on earthquake tremors. For this reason, the Japan Meteorological Agency issues tsunami information and predicted values, and residents evacuate based on this information. This is why tsunami information is extremely important. Up to the present, higher speed has been achieved in monitoring and prediction, as seen in the JMA's Earthquake Early Warning (EEW) system. In Japan, tsunami information can now be issued approximately three minutes after an earthquake. Although this information system is the most advanced of its type in the world, challenges still remain on both the information providing side and the information receiving side.

First, one issue on the information providing side is how far it is possible to issue detailed, high accuracy information. At present, quantitative information is issued by dividing the whole country into 66 districts. However, considering the local nature of tsunamis and the response in the area, this is not adequate. There are also phenomena that cannot be predicted by the standard techniques. For example, there are tsunami earthquakes that cause large tsunamis in spite of producing only small tremors, and there are non-seismic tsunamis which attack even when no tremors occur in the area, as exemplified by remote tsunamis. In order to increase accuracy and reliability in predicting these phenomena, it is important to monitor actual tsunamis further out to sea, and to use the results to correct the predicted values. Thus, high expectations are placed on the existing monitoring data networks in coastal regions, GPS tsunami-monitoring buoys, and similar systems.

On the other hand, there are two issues on the information receiving side. First, there are questions as to whether the need to evacuate can be judged, and coastal damage can be predicted, from the present tsunami information (height and arrival time of the tsunami). Damage does not necessarily occur simply because a tsunami is large, and conversely, even small tsunamis sometimes cause serious damage. Damage can be viewed as the result of the discrepancy between the magnitude of an external force, that is, a tsunami, and the response capabilities for preventing damage by that force, or disaster prevention capabilities. In other words, damage cannot be judged from tsunami information which concerns only external force. It is necessary to fill the gap between the present tsunami information and damage. For this, an evaluation of disaster prevention capabilities is necessary. Furthermore, because there is no time margin to confirm disaster prevention capabilities after an earthquake occurred, preparations must be made in advance. Regional inundation maps, disaster prevention maps, and similar tools are necessary information for bridging this gap. The second issue on the information receiving side is the strong tendency of each individual to make judgments about tsunamis based on his or her own personal standards. Even if various information is provided, people judge for themselves that they are safe, and as a result, they fail to take action to avoid disaster. Thus, it is necessary to understand information about tsunamis correctly, and to have a knowledge of the correct response for disaster mitigation. Each individual must be able to understand the conditions that are occurring, and must heighten his or her ability to select the proper response corresponding to those conditions.

Epilogue

– To Those Who Have Finished Reading This Book –

This book was written by leading Japanese tsunami researchers with the aim of saving all lives in tsunamis. The basic rules of tsunami survival are simple: Make everyday preparations, evacuate immediately when a tsunami attacks, and move as quickly as possible to high ground. We believe that everyone who has finished reading the manual understands these points. Therefore, we would like to make the following two requests to our readers.

You now understand the experience of a tsunami, and you have gained much knowledge about this natural phenomenon. However, you will probably never experience an actual tsunami. To put your experience and knowledge to good use, we ask that you communicate what you have learned to those around you. It is not necessary to create special occasions for this. Simply mention it from time to time when you have the chance. In particular, it is important to educate the younger generation, that is, students and young children. Because tsunamis are so rare, many regions will not experience a tsunami once in a century. Over time, we all forget the tragedy that tsunamis can cause. In Japan, a noted physicist expressed this by saying "disaster come when we forget." Therefore, we hope that you will discuss this with children so that future generations will also be aware of tsunamis, and will know how to react if one occurs.

We also have another difficult request. When a tsunami warning is announced and people are evacuating, it is natural to feel that you must help children, older people, and persons with handicaps. It may sound cold-blooded, but your own safety must be your first consideration. In other words, you can't help others if you are struggling to save your own life. Ensure your own safety, then help others. For example, Figure shows part of the emergency guide carried on airlines. You have probably heard the announcement in a plane before takeoff: "When the oxygen masks are released, put on your own mask first. Do not attempt to assist children or others until you have put on your own mask." The same rule applies in tsunami evacuations. Keep a cool head, first ensuring your own safety, and then act in response to the changing circumstances.

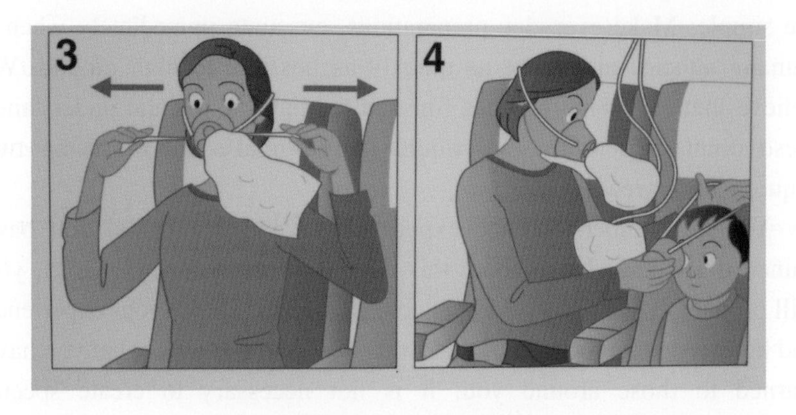

Figure Airline Safety Guide (JAL)

Index